Video Basics 2 Workbook

Herbert Zettl

San Francisco State University

Wadsworth Publishing Company
I(T)P® An International Thomson Publishing Company

Belmont, CA ▪ Albany, NY ▪ Bonn ▪ Boston ▪ Cincinnati ▪ Detroit ▪ Johannesburg ▪ London ▪ Madrid
Melbourne ▪ Mexico City ▪ New York ▪ Paris ▪ Singapore ▪ Tokyo ▪ Toronto ▪ Washington

Communications Editor: Randall Adams
Assistant Editor: Michael Gillespie
Editorial Assistant: Megan Gilbert
Marketing Manager: Mike Dew
Project Editor: Vicki Friedberg
Copy Editor: Elizabeth Von Radics
Production: Ideas to Images
Designer and Art Director: Gary Palmatier
Print Buyer: Barbara Britton
Permissions Editor: Robert M. Kauser
Technical Illustrator: Robaire Ream, Ideas to Images
Cover Designer: Gary Palmatier, Ideas to Images
Cover Photographers: Robert Brenner, Gary Palmatier
Compositor: Ideas to Images
Printer: Malloy Lithographing

Photo Credits
Daniel Hubbell: 28 (nos. 17, 21), 69 (no. 60)
The Grass Valley Group: 104, 106
Ideas to Images: 79 (ex. 5)
Lowel-Light Mfg., Inc.: 52 (nos. 54, 58)
Mole-Richardson Co.: 52 (nos. 51, 53, 56, 57, 59)
Steve Renick: 115, 116, 117, 118
Selco Products Company: 79 (ex. 4)
Alex Zettl: 68 (d. top)
Herbert Zettl: all other photos

The audio signature used on the back cover and title page of this book represents the spoken words *Video Basics* as they are digitally represented in Adobe Premiere, a nonlinear video-editing program.

Copyright © 1998 by Wadsworth Publishing Company

A Division of International Thomson Publishing Inc.
I(T)P® The ITP logo is a registered trademark under license.

Printed in the United States of America

2   3   4   5   6   7   8   9   10

For more information, contact Wadsworth Publishing Company, 10 Davis Drive, Belmont, CA 94002, or electronically at http://www.thomson.com/wadsworth.html

International Thomson Publishing Europe
Berkshire House 168-173
High Holborn
London, WC1V 7AA, England

Thomas Nelson Australia
102 Dodds Street
South Melbourne 3205
Victoria, Australia

Nelson Canada
1120 Birchmount Road
Scarborough, Ontario
Canada M1K 5G4

International Thomson Publishing GmbH
Königswinterer Strasse 418
53227 Bonn, Germany

International Thomson Editores
Campos Eliseos 385, Piso 7
Col. Polanco
11560 México D.F. México

International Thomson Publishing Asia
221 Henderson Road
#05-10 Henderson Building
Singapore 0315

International Thomson Publishing Japan
Hirakawacho Kyowa Building, 3F
2-2-1 Hirakawacho
Chiyoda-ku, Tokyo 102, Japan

International Thomson Publishing Southern Africa
Building 18, Constantia Park
240 Old Pretoria Road
Halfway House, 1685 South Africa

All rights reserved. No part of this work covered by the copyright hereon may be reproduced or used in any form or by any means—graphic, electronic, or mechanical, including photocopying, recording, taping, or information storage and retrieval systems—without the written permission of the publisher.

ISBN 0-534-52687-X

Preface

Whatever you learn and retain in your memory is of little use if you can't recall it on command and apply it effectively to a given problem-solving task. Whereas the *Video Basics 2* text is designed to function primarily as an input and storage device (learning), *Video Basics 2 Workbook* is meant to reinforce your learning (input and storage)—and especially aid your retrieval (recall) and effective use (application) of information. Its main purpose is to reinforce your learning and help you bridge the gap between reading and doing. *Zettl's Video Lab 2.0* CD-ROM (ZVL 2.0) further facilitates this difficult jump from reading to doing. Here are some tips on how to use the *Workbook*:

▶ *To reinforce your input—the learning of video terminology and production tools, procedures, and techniques.* The review of key terms and the multiple-choice questions are designed to test your knowledge of terminology and the basic tools of video production—what they are and how they work. These exercises are to check your learning. If you have no trouble matching the key terms with their proper definitions and can answer the multiple-choice questions with relative ease without having to look up the answers in *Video Basics 2*, you have successfully learned the required material. By using *ZVL 2.0*, you can test yourself again on the accuracy of your recall and, especially, on how to apply the various production tasks.

▶ *To facilitate retrieval—the recall of information when needed.* You can use the *Workbook* diagnostically to find out what you know and what you don't, even after reading *Video Basics 2* or relying on your actual video production experience. By filling out the review sections (except the problem-solving applications) without looking up any answers, you will quickly discover your strong and weak points. After pinpointing your strengths and weaknesses, you can go back to the text and *ZVL 2.0* and remedy your deficiencies before engaging in actual production practices.

▶ *To facilitate the efficient and effective application of video production equipment and practices.* The multiple-choice questions test you on what you should do in various production situations—and how to do it. *ZVL 2.0* gives you further opportunities to practice similar applications. For especially relevant practice sessions, see the various ZVL references in the *Video Basics 2* text.

▶ *To find creative solutions to production problems.* The problem-solving applications challenge you to come up with creative solutions to common production problems. Note that each problem-solving section tolerates various answers, depending on the specific production context or available equipment. Let your mind wander and don't feel limited by budget and time restrictions. You can use the floor plan and storyboard sheets both to solve some of the assigned *Workbook* problems and for your own video productions that are independent of the text requirements.

Video Basics 2 Workbook also gives your instructor an objective tool with which to assess your learning, retrieval, and application skills. When using *ZVL 2.0*, you can record your quiz answers on a floppy disk that you can submit to your instructor. Make sure that you put your name on the diskette, have named it *ZVL*, and that the disk is not write-protected when using it.

Once again, my first thanks go to the people at Wadsworth Publishing Company who insisted on a workbook that is as efficient in its use as it is effective in aiding student learning. I am indebted also to my colleagues at San Francisco State University and other institutions, and especially to my students, who helped me directly or indirectly by asking questions, by making mistakes I expected, and by finding solutions I did not expect. Special thanks go to my wife, Erika, who as a longtime classroom teacher, administrator, and educational consultant once again helped me objectify the answers without impinging on the students' creativity.

Herbert Zettl

Contents

Part I: **Production Processes and People**   1

Chapter 1   **The Production Process**   3
Review of Key Terms   3
Review of Effect-to-Cause Production Model   4
Review Quiz   6
Zettl's Video Lab 2.0 Quiz   7
Problem-Solving Applications   7

Chapter 2   **The Production Team: Who Does What When?**   9
Review of Key Terms   9
Review of Production Personnel and Responsibilities   11
Review of the Production Schedule   12
Review Quiz   13
Zettl's Video Lab 2.0 Quiz   14
Problem-Solving Applications   14

Part II: **Image Creation and Control**   15

Chapter 3   **The Video Camera**   17
Review of Key Terms   17
Review of Basic Image Formation   19
Review of Camera Function and Elements   20
Review of Lenses   21
Review of the Imaging Device and Camera Chain   22
Review Quiz   23
Zettl's Video Lab 2.0 Quiz   24
Problem-Solving Applications   24

Chapter 4   **Looking Through the Viewfinder**   25
Review of Key Terms   25
Review of Framing a Shot and Picture Composition   28
Review of Vectors and Psychological Closure   32
Review of Lenses, Depth of Field, and Z-axis Manipulation   34
Review Quiz   36
Zettl's Video Lab 2.0 Quiz   37
Problem-Solving Applications   37

Chapter 5 **Operating the Camera**   39
   Review of Key Terms   39
   Review of Camera Mounts   41
   Review of Operational Features   44
   Review Quiz   45
   Zettl's Video Lab 2.0 Quiz   46
   Problem-Solving Applications   46

Chapter 6 **Light, Color, Lighting**   47
   Review of Key Terms   47
   Review of Light, Shadows, Color   51
   Review of Lighting Instruments   52
   Review of Lighting Techniques   55
   Review Quiz   58
   Zettl's Video Lab 2.0 Quiz   59
   Problem-Solving Applications   59

Chapter 7 **Visual Effects and Computer-Generated Video**   61
   Review of Computer Terminology   61
   Review of the Desktop Computer System   64
   Review of Key Terms   65
   Review of Standard Electronic Effects   66
   Review Quiz   70
   Zettl's Video Lab 2.0 Quiz   71
   Problem-Solving Applications   71

Chapter 8 **Audio and Sound Control**   73
   Review of Key Terms   73
   Review of Sound-Generating Elements and Sound Pickup   77
   Review of Microphone Use   78
   Review of Sound Control   79
   Review of Sound Recording and Aesthetics   80
   Review Quiz   81
   Zettl's Video Lab 2.0 Quiz   82
   Problem-Solving Applications   82

Part III: **Video Recording, Storage, and Sequencing**   83

Chapter 9 **Video Recording**   85
   Review of Key Terms   85
   Review of Videotape Recording Systems   88
   Review of the Video-Recording Process   89
   Review of Nonlinear Storage Systems   89
   Review of Interactive Video and Multimedia   90
   Review Quiz   91
   Zettl's Video Lab 2.0 Quiz   92
   Problem-Solving Applications   92

Chapter 10 **Editing Principles**   93
  Review of Key Terms 93
  Review of Aesthetic Principles of Continuity Editing 95
  Review Quiz 99
  Zettl's Video Lab 2.0 Quiz 100
  Problem-Solving Applications 100

Chapter 11 **Switching and Postproduction Editing**   101
  Review of Key Terms 101
  Review of Basic Switcher Operation 104
  Review of Postproduction Editing 107
  Review Quiz 109
  Zettl's Video Lab 2.0 Quiz 110
  Problem-Solving Applications 110

Part IV: Talent and the Production Environment   111

Chapter 12 **Talent, Clothing, and Makeup**   113
  Review of Key Terms 113
  Review of Performing Techniques 115
  Review of Acting Techniques 123
  Review of Clothing and Makeup 125
  Review Quiz 126
  Zettl's Video Lab 2.0 Quiz 127
  Problem-Solving Applications 127

Chapter 13 **Production Environment: The Studio**   129
  Review of Key Terms 129
  Review of Video Production Studio 131
  Review of Scenery, Properties, and Scenic Design 132
  Review Quiz 136
  Problem-Solving Applications 137

Chapter 14 **Field Production and Synthetic Environments**   141
  Review of Key Terms 141
  Review of Remotes and Field Productions 143
  Review of Synthetic Environments 147
  Review Quiz 148
  Zettl's Video Lab 2.0 Quiz 149
  Problem-Solving Applications 149

Floorplan Grids and Storyboard Forms   151

Part I

Production Processes and People

Course No. _____ Date _____ Name _____

The Production Process

REVIEW OF KEY TERMS

Match each term with its appropriate definition by filling in the corresponding bubble.

1. preproduction
2. production
3. postproduction
4. process message
5. medium requirements
6. defined process message

A. The message actually received by the viewer in the process of watching a video program.

B. Preparation of all production details.

C. All personnel, equipment, and facilities needed for a production, as well as budgets, schedules, and the various production phases.

D. Any production activity that occurs after the production. Usually refers to either videotape editing or audio sweetening.

E. The desired effect of the program on the viewer.

F. The actual activities in which an event is videotaped and/or televised.

SECTION TOTAL

Chapter 1 — The Production Process

REVIEW OF EFFECT-TO-CAUSE PRODUCTION MODEL

1. Identify each part of the effect-to-cause production diagram, and fill in the bubbles with the corresponding numbers.

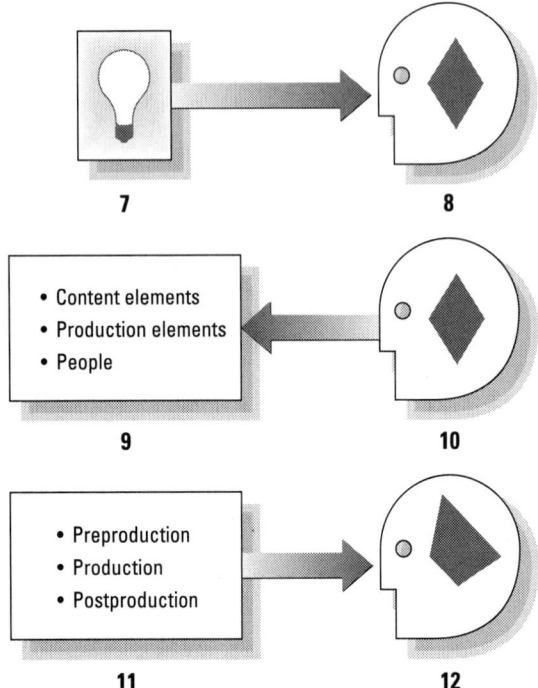

a. defined process message [appears twice]

b. actual process message (effect)

c. basic idea

d. medium requirements (cause)

1a ○7 ○8 ○9
   ○10 ○11 ○12

1b ○7 ○8 ○9
   ○10 ○11 ○12

1c ○7 ○8 ○9
   ○10 ○11 ○12

1d ○7 ○8 ○9
   ○10 ○11 ○12

PAGE TOTAL [   ]

Part I — Production Processes and People

Course No. _____ Date _____ Name _____

Select the correct answers, and fill in the bubbles with the corresponding numbers.

2. The message that ultimately counts is the one that is (13) *carefully constructed by the producer* (14) *transmitted by the originating institution* (15) *received and interpreted by the viewer.*

3. This message is called the (16) *optimal message* (17) *actual process message* (18) *goal-directed message.*

4. The medium requirements include (19) *equipment, but not program content and people* (20) *equipment and people* (21) *equipment, production elements, and people.*

5. The most important initial step in the effect-to-cause model is to (22) *determine the available production equipment* (23) *define the process message* (24) *contact production personnel.*

6. Medium requirements are basically determined by (25) *the process message* (26) *the chief engineer* (27) *the available equipment.*

Chapter 1 — *The Production Process*

REVIEW QUIZ

*Mark the following statements as true or false by filling in the bubbles in the **T** (for true) or **F** (for false) column.*

1. Because production is primarily a creative activity, any type of production system would prove counterproductive.

2. The process message is the message actually received by the viewer.

3. Clustering and brainstorming are similar idea-creating techniques.

4. Medium requirements include equipment and facilities, but not production personnel.

5. Formative evaluation cannot be done until the program is finished.

6. The audience is relatively unimportant when using the effect-to-cause production model.

7. The closer the actual and defined process messages match, the more successful the communication is.

8. When brainstorming for new ideas, somebody should make sure that the ideas generated are relevant to the topic.

9. If you plan on doing extensive postproduction, you do not have to engage in preproduction.

10. A precise description of a specific target audience should be based on demographic as well as psychographic data.

Course No. _____ Date _____ Name _____

ZETTL'S VIDEO LAB 2.0 QUIZ

*Click on the **process** monitor and take the quiz on tape 4 **Ideas**.*

PROBLEM-SOLVING APPLICATIONS

1. Do a brainstorming session. Place several people engaged in the same production problem in a circle, and place the mic of a small audiotape recorder in the center of the circle. Describe the general theme of the production, such as "doing something about the homeless and the hungry in this country" or "how to get high-school students interested in math." Do not try to state the general idea as a process message. The specific process message should develop out of this brainstorming session.

    A good way to begin is to have one of the group members say something neutral, such as "Knock-knock—who's there?" or "Hello, what can I do for you?" or something of that order. Make sure that *all* ideas are accepted and not commented on, however far-out or ridiculous they may be. During the playback, you can be more discriminating and select only the ideas that fit your overall production theme. See whether, and how, this material might help you design a process message and/or the medium translation of this process message.

2. Select a key word signifying your program theme and do an idea cluster. Use the cluster to develop the process message and the specific program (interview, documentary, drama) and production approach (studio, ENG, EFP).

3. Expand the three clusters indicated in the following three figures and develop a precise process message from each of them.

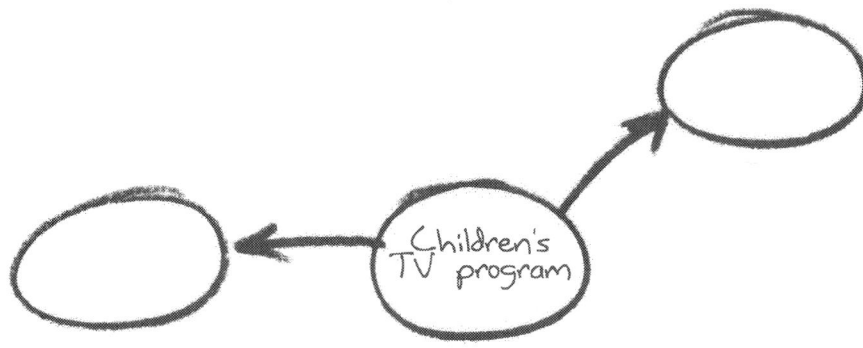

Chapter 1 — *The Production Process*

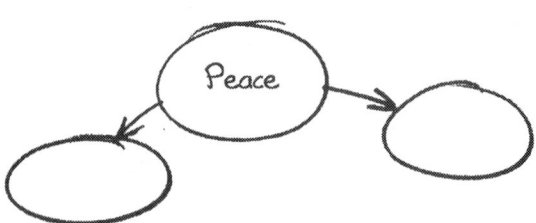

2 The Production Team: Who Does What When?

REVIEW OF KEY TERMS

Match each term with its appropriate definition by filling in the corresponding bubble.

1. above-the-line
2. below-the-line
3. production schedule
4. preproduction team
5. production team
6. postproduction team
7. floor plan

A. A diagram of scenery, properties, and set dressings drawn on a grid pattern.

B. Normally consists of the video editor and, for complex productions, a sound designer who remixes the sound track.

C. Consists of people who plan the production. This team normally includes the producer, director, writer, art director, and technical supervisor or TD. Large productions may also include a composer and a choreographer.

D. Category for technical personnel, including such crew members as camera operators, floor persons, and audio engineers.

E. Consists of a variety of nontechnical and technical people, such as producer and various assistants (associate producer and PA), the director and assistant (AD), and the talent and production crew.

1. above-the-line
2. below-the-line
3. production schedule
4. preproduction team
5. production team
6. postproduction team
7. floor plan

F. A listing of the start times of major production events.

G. Category for nontechnical personnel, such as producers, directors, and talent.

Course No. _____ Date _____ Name _____

REVIEW OF PRODUCTION PERSONNEL AND RESPONSIBILITIES

Identify the production person mainly responsible for the following production situations, and fill in the bubbles with the corresponding numbers.

1. To increase the overall light level in a scene, you should ask (8) *the LD* (9) *the PA* (10) *the floor manager.*

2. The producer complains about the "static look" of the show. To correct the situation, she must talk to (11) *the TD* (12) *the art director* (13) *the executive producer.*

3. The talent wants to make sure that he can see the opening cues. He should talk to (14) *the director* (15) *the producer* (16) *the floor manager.*

4. Informing talent and crew about a videotaping schedule change is the responsibility of (17) *the floor manager* (18) *the producer* (19) *the director.*

5. To make sure that the dancers can hear the music is the responsibility of (20) *the LD* (21) *the AD* (22) *the audio engineer.*

6. Staying in focus during a long camera movement is done by (23) *the camera operator* (24) *the video operator* (25) *the director.*

7. To have the scenery set up and the set decorated by the scheduled time is the responsibility of (26) *the floor manager* (27) *the director* (28) *the TD.*

8. The set design is done by (29) *the floor manager* (30) *the director* (31) *the art director.*

9. Last-minute changes to the names on the final credits are made by (32) *the art director* (33) *the TD* (34) *the C.G. operator.*

10. The person in charge of camera rehearsals is (35) *the director* (36) *the AD* (37) *the TD.*

SECTION TOTAL

Chapter 2 — *The Production Team: Who Does What When?*

REVIEW OF THE PRODUCTION SCHEDULE

Analyze the following production schedule (time line) for major omissions. From the list below, select the major items omitted from the production schedule, and fill in the corresponding bubbles.

Production Schedule April 15: Panel Discussion (Studio 1)

8:30–9:00 A.M.	Tech meeting
9:00–11:00 A.M.	Setup and lighting
12:00–12:15 P.M.	Notes and reset
12:15–12:30 P.M.	Briefing of panel guests
12:30–12:45 P.M.	Run-through and camera rehearsal
12:45–12:55 P.M.	Notes and reset
12:55–2:55 P.M.	Taping
2:55–3:30 P.M.	Spill

(38) *tech meeting*

(39) *taping*

(40) *crew call*

(41) *notes*

(42) *strike*

(43) *camera rehearsal*

(44) *break*

(45) *lighting*

(46) *reset*

(47) *meal*

(48) *budget meeting*

(49) *transportation from and to airport*

Course No. _____ Date _____ Name _____

REVIEW QUIZ

*Mark the following statements as true or false by filling in the bubbles in the **T** (for true) or **F** (for false) column.*

		T	F
1.	The floor manager is principally responsible for the budget.	1 ○ 50	○ 51
2.	The audio engineer works the audio console during a show.	2 ○ 52	○ 53
3.	Talent includes actors but not performers.	3 ○ 54	○ 55
4.	The LD is in charge of directing the log entries.	4 ○ 56	○ 57
5.	The PA is responsible for notes.	5 ○ 58	○ 59
6.	The executive producer is always part of the field survey team.	6 ○ 60	○ 61
7.	Normally, the TD runs the VTRs.	7 ○ 62	○ 63
8.	Principal camera positions and talent blocking are determined by the director.	8 ○ 64	○ 65
9.	The art director is responsible for putting up the studio set.	9 ○ 66	○ 67
10.	The production and postproduction teams are identical.	10 ○ 68	○ 69

SECTION TOTAL [ ]

Chapter 2 — The Production Team: Who Does What When?

ZETTL'S VIDEO LAB 2.0 QUIZ

Click on the **process** monitor and take the quiz on tape 2 **Phases**.

PROBLEM-SOLVING APPLICATIONS

1. The camera operators for a weekly two-camera interview series tell you, the producer, that they do not need a director because they have done the show many times and know every shot by heart. Do you agree? If so, why? If not, why not?

2. The ENG/EFP camcorder operator of a weekly on-location interview show tells you, the producer, that he does not need a director because he has done the show many times and knows what shots are required. Do you agree? If so, why? If not, why not?

3. The video manager of a corporation tells you that she has agreed to pay specific fees for the above-the-line personnel, but a lump sum for all below-the-line costs. What does she mean?

4. The PA complains that the producer, the director, and even the floor manager ask her to write down specific production problems. Her comment is that, as a production assistant, she has to listen only to the producer. Is her complaint justified?

5. The talent complains to the director about the long hours and the relatively low pay. Is the talent complaining to the right person? If so, why? If not, to whom should the talent direct the complaints?

Part II

Image Creation and Control

Course No. _____ Date _____ Name _____

3 The Video Camera

REVIEW OF KEY TERMS

Match each term with its appropriate definition by filling in the corresponding bubble.

1. camera chain
2. camcorder
3. CCD
4. ENG/EFP camera
5. HDTV camera
6. field
7. frame
8. aperture
9. focal length
10. zoom range
11. *f*-stop
12. beam splitter

A. A portable camera with the VTR attached to it to form a single, independent unit.

B. An electronic "chip." It is the most common imaging device in color cameras.

C. The camera connected with the CCU, power supply, and sync generator.

D. A complete scanning cycle, consisting of two fields.

PAGE TOTAL

1. camera chain
2. camcorder
3. CCD
4. ENG/EFP camera
5. HDTV camera
6. field
7. frame
8. aperture
9. focal length
10. zoom range
11. *f*-stop
12. beam splitter

E. The scanning of all odd- *or* even-numbered lines that occurs every 1/60 second.

F. A special, high-resolution camera.

G. The iris opening of a lens.

H. The calibration on the lens indicating the aperture (and therefore the amount of light passing through the lens).

I. The degree to which the focal length can be changed from a wide shot to a close-up during a zoom.

J. A portable camera, without a built-in VTR, that contains all camera controls in the camera itself.

Part II — Image Creation and Control

Course No. _____ Date _____ Name _____

1. camera chain	5. HDTV camera	9. focal length
2. camcorder	6. field	10. zoom range
3. CCD	7. frame	11. *f*-stop
4. ENG/EFP camera	8. aperture	12. beam splitter

K. Optical device within the camera that splits the white light into three primary colors: red, green, and blue.

L. How much of a scene the lens can see and how magnified the distant object looks.

K ○ ○ ○ ○
  1 2 3 4
  ○ ○ ○ ○
  5 6 7 8
  ○ ○ ○ ○
  9 10 11 12

L ○ ○ ○ ○
  1 2 3 4
  ○ ○ ○ ○
  5 6 7 8
  ○ ○ ○ ○
  9 10 11 12

PAGE TOTAL ☐

SECTION TOTAL ☐

REVIEW OF BASIC IMAGE FORMATION

Select the correct answers, and fill in the bubbles with the corresponding numbers.

1. A television frame consists of (13) *four fields* (14) *two fields* (15) *three fields*.

2. The scanning of a single field takes (16) *1/20 second* (17) *1/30 second* (18) *1/60 second*.

3. The scanning of a complete frame takes (19) *1/20 second* (20) *1/30 second* (21) *1/60 second*.

1 ○   ○   ○
  13   14   15

2 ○   ○   ○
  16   17   18

3 ○   ○   ○
  19   20   21

SECTION TOTAL ☐

Chapter 3 — The Video Camera

REVIEW OF CAMERA FUNCTION AND ELEMENTS

Select the correct answers, and fill in the bubbles with the corresponding numbers.

1. The three basic parts of the television camera are (22) *pedestal* (23) *lens* (24) *VTR* (25) *camera* (26) *viewfinder* (27) *tally light.*

2. Fill in the bubbles whose numbers correspond with the camera elements shown in the following figure.

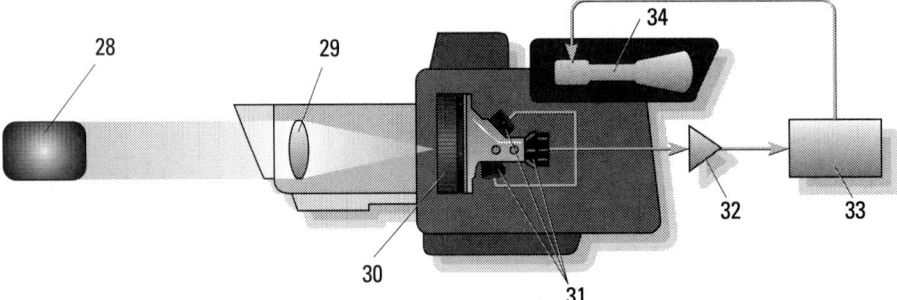

   a. Transforms light into electric energy or video signals.

   b. Converts video signals back into visible screen images.

   c. Reflects light.

   d. Amplifies video signals.

   e. Processes video signal.

   f. Gathers and transmits the light.

   g. Splits the white light into red, green, and blue light beams.

Part II — *Image Creation and Control*

Course No. _____ Date _____ Name _____

REVIEW OF LENSES

Select the correct answers, and fill in the bubbles with the corresponding numbers.

1. A fast lens transmits an image (35) *faster* (36) *slower* than a slow lens, or permits (37) *more* (38) *less* light to enter, assuming a maximum aperture.

   1 ○ ○ ○ ○
       35 36 37 38

2. A slow lens transmits an image (39) *faster* (40) *slower* than a fast lens, or permits (41) *more* (42) *less* light to enter, assuming a maximum aperture.

   2 ○ ○ ○ ○
       39 40 41 42

3. In the diagram below, select the most appropriate *f*-stop number for each of the four apertures (a through d), and fill in the bubbles with the corresponding number.

   a. 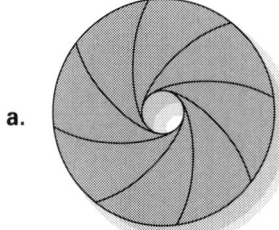   (43) *f*/22 (44) *f*/5.6 (45) *f*1.4

   3a ○   ○   ○
       43   44   45

   b.    (46) *f*/1.4 (47) *f*/2.8 (48) *f*/16

   3b ○   ○   ○
       46   47   48

   c. 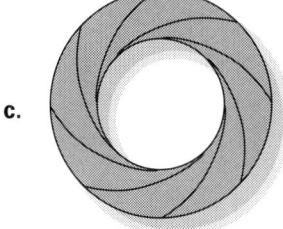   (49) *f*/1.4 (50) *f*/4 (51) *f*/16

   3c ○   ○   ○
       49   50   51

   d. 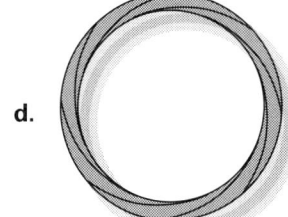   (52) *f*/1.4 (53) *f*/8 (54) *f*/22

   3d ○   ○   ○
       52   53   54

   SECTION TOTAL ☐

Chapter 3 — *The Video Camera*

REVIEW OF THE IMAGING DEVICE AND CAMERA CHAIN

Select the correct answers, and fill in the bubbles with the corresponding numbers.

1. The more pixels a CCD imaging device contains, the (55) *sharper* (56) *brighter* (57) *more colorful* the resulting screen image will be.

2. The highest-quality images are produced by (58) *single-chip* (59) *two-chip* (60) *three-chip* cameras.

3. Camcorders do not need a (61) *power supply* (62) *VTR* (63) *cable that connects them to the VTR*.

Course No. _____ Date _____ Name _____

REVIEW QUIZ

Mark the following statements as true or false by filling in the bubbles in the T (for true) or F (for false) column.

		T	F
1.	The standard multicore camera cables can be used over longer distances from camera to CCU than triax and fiber-optic cables.	1 ○ 64	○ 65
2.	A CCD translates light into electric energy.	2 ○ 66	○ 67
3.	The CCU performs camera setup and control functions, but does not help with keeping in focus during a zoom.	3 ○ 68	○ 69
4.	The sync generator and power supply fulfill similar functions.	4 ○ 70	○ 71
5.	The terms *ENG/EFP cameras* and *camcorders* are identical.	5 ○ 72	○ 73
6.	Camcorders can come with or without a VTR.	6 ○ 74	○ 75
7.	A beam splitter divides the incoming white light into red, green, and blue light beams.	7 ○ 76	○ 77
8.	The ENG/EFP camera contains the major parts of the regular camera chain.	8 ○ 78	○ 79
9.	A complete television frame is scanned every 1/60 second.	9 ○ 80	○ 81
10.	A television frame and a scanning field are the same.	10 ○ 82	○ 83
11.	The wide-angle position of the zoom lens provides a vista similar to that of a short-focal-length lens.	11 ○ 84	○ 85
12.	The lower the *f*-stop number, the larger the aperture.	12 ○ 86	○ 87
13.	ENG/EFP cameras can be connected to an RCU.	13 ○ 88	○ 89
14.	Zoom range refers to how fast you can zoom in or out.	14 ○ 90	○ 91
15.	A composite (NTSC) signal combines the color (C) and luminance (Y) signals.	15 ○ 92	○ 93

SECTION TOTAL

Chapter 3 — The Video Camera

ZETTL'S VIDEO LAB 2.0 QUIZ

Click on the **camera** monitor and take the quizzes on tape 2 **Zoom Lens**, tape 3 **Exposure Control**, and tape 4 **Focusing**.

PROBLEM-SOLVING APPLICATIONS

1. The camera operator tells you not to worry about the bright white cap of the dark-skinned golf champion. He says the automatic iris of the camcorder will take care of the proper exposure. What is your reaction?

2. The director of a field production is worried that the cameras might not deliver pictures whose colors match when edited together. The TD tells the director that the use of RCUs will greatly help in color matching. Do you agree with the TD?

3. The director tells you that for covering an outdoor sporting event, a lens with a great zoom range is more important than an extremely fast one. Why does the director think so?

Course No. _____ Date _____ Name _____

4 Looking Through the Viewfinder

REVIEW OF KEY TERMS

Match each term with its appropriate definition by filling in the corresponding bubble.

1. psychological closure
2. headroom
3. noseroom
4. depth of field
5. field of view
6. z-axis
7. close-up
8. vector
9. medium shot
10. long shot
11. cross-shot
12. over-the-shoulder shot

A. Object seen from far away or framed very loosely.

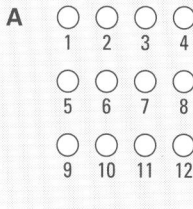

B. The portion of a scene visible through a particular lens; its vista.

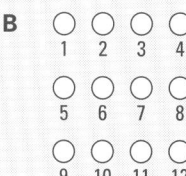

C. Object or any part of it seen at close range.

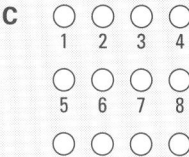

D. Similar to over-the-shoulder shot, except that the camera-near person is completely out of the shot.

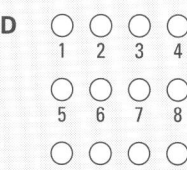

PAGE TOTAL

Chapter 4 — *Looking Through the Viewfinder*

1. psychological closure
2. headroom
3. noseroom
4. depth of field
5. field of view
6. z-axis
7. close-up
8. vector
9. medium shot
10. long shot
11. cross-shot
12. over-the-shoulder shot

E. Mentally filling in missing visual information that will lead to a complete and stable configuration.

F. Object seen from a medium distance.

G. The space left in front of a person looking toward the edge of the screen.

H. The space left between the top of the head and the upper screen edge.

I. Indicates screen depth. Extends from camera lens to horizon.

J. Camera looks over the camera-near person's shoulder.

Course No. _____ Date _____ Name _____

1. psychological closure	5. field of view	9. medium shot
2. headroom	6. z-axis	10. long shot
3. noseroom	7. close-up	11. cross-shot
4. depth of field	8. vector	12. over-the-shoulder shot

K. The area in which all objects, located at different distances from the camera, are in focus.

L. A directional screen force.

Chapter 4 — *Looking Through the Viewfinder*

REVIEW OF FRAMING A SHOT AND PICTURE COMPOSITION

1. Using the following set of numbered images, fill in the corresponding bubble for each of the fields of view or other shot designations listed below.

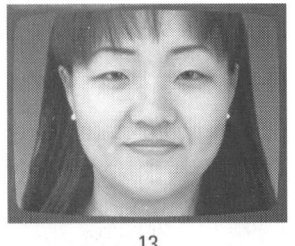
13

14

15

16

17

18

19

20

21

    a. ELS (extreme long shot)

    b. LS (long shot)

    c. MS (medium shot)

Part II — Image Creation and Control

Course No. _____ Date _____ Name _____

d. CU (close-up)

1d ○ ○ ○ ○ ○
    13 14 15 16 17
    ○ ○ ○ ○
    18 19 20 21

e. ECU (extreme close-up)

1e ○ ○ ○ ○ ○
    13 14 15 16 17
    ○ ○ ○ ○
    18 19 20 21

f. three-shot

1f ○ ○ ○ ○ ○
    13 14 15 16 17
    ○ ○ ○ ○
    18 19 20 21

g. knee shot

1g ○ ○ ○ ○ ○
    13 14 15 16 17
    ○ ○ ○ ○
    18 19 20 21

h. bust shot

1h ○ ○ ○ ○ ○
    13 14 15 16 17
    ○ ○ ○ ○
    18 19 20 21

i. over-the-shoulder shot

1i ○ ○ ○ ○ ○
    13 14 15 16 17
    ○ ○ ○ ○
    18 19 20 21

PAGE TOTAL ☐

Chapter 4 — *Looking Through the Viewfinder*

2. Evaluate the framing of shots in the next six figures by filling in the bubbles with the corresponding numbers.

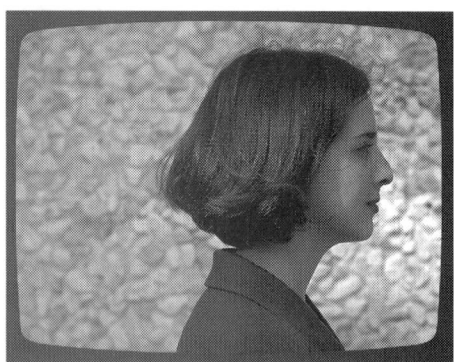

a. This shot is (22) *acceptable* (23) *unacceptable* because it has (24) *too much noseroom* (25) *too little noseroom* (26) *too much headroom* (27) *too much leadroom*.

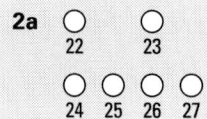

b. This CU is (28) *acceptable* (29) *unacceptable* because it has (30) *no headroom* (31) *too much headroom* (32) *no noseroom* (33) *sufficient clues for closure in off-screen space*.

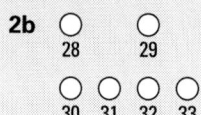

c. This ECU is (34) *acceptable* (35) *unacceptable* because it has (36) *no headroom* (37) *no leadroom* (38) *sufficient clues for closure in off-screen space* (39) *no noseroom*.

Part II — Image Creation and Control

Course No. _____ Date _____ Name _____

d. This over-the-shoulder shot is (40) *acceptable* (41) *unacceptable*.

2d ○ ○
   40 41

e. This shot is intended to emphasize the imposing nature of big city buildings. Its framing is (42) *acceptable* (43) *unacceptable*.

2e ○ ○
   42 43

f. This shot makes (44) *good* (45) *poor* use of screen depth. If depth improvement is needed, you should (46) *add foreground objects* (47) *take a tighter shot of the mountains*.

2f ○ ○
   44 45
   ○ ○
   46 47

PAGE TOTAL ☐

SECTION TOTAL ☐

Chapter 4 — Looking Through the Viewfinder

REVIEW OF VECTORS AND PSYCHOLOGICAL CLOSURE

1. Identify the specific vectors displayed in the following three figures by filling in the bubbles with the corresponding numbers.

a. This picture shows (48) *graphic* (49) *index* (50) *motion* vectors.

b. This picture shows prominent (51) *graphic* (52) *index* (53) *motion* vectors.

c. This picture shows prominent (54) *graphic* (55) *index* (56) *motion* vectors.

Part II — Image Creation and Control

Course No. _____ Date _____ Name _____

2. Evaluate the following two figures by filling in the bubbles with the corresponding numbers.

a. The framing of this shot is (57) *acceptable* (58) *unacceptable* because it (59) *leads us to undesirable closure within the frame* (60) *leads us to undesirable closure in off-screen space* (61) *leads us to desirable closure in off-screen space* (62) *prevents closure.*

2a ○ ○
   57 58
  ○ ○ ○ ○
  59 60 61 62

b. The framing of this shot is (63) *acceptable* (64) *unacceptable* because it (65) *leads us to undesirable closure within the frame* (66) *leads us to undesirable closure in off-screen space* (67) *leads us to desirable closure in off-screen space* (68) *prevents closure.*

2b ○ ○
   63 64
  ○ ○ ○ ○
  65 66 67 68

PAGE TOTAL [ ]

SECTION TOTAL [ ]

Chapter 4 — *Looking Through the Viewfinder*

REVIEW OF LENSES, DEPTH OF FIELD, AND Z-AXIS MANIPULATION

Select the correct answers, and fill in the bubbles with the corresponding numbers.

1. The area in which all objects, although located at different distances from the camera, are in focus is called (69) *depth of focus* (70) *depth of focal length* (71) *depth of field.*

2. A wide-angle lens position gives you a relatively (72) *shallow* (73) *wide* (74) *great* depth of field.

3. A narrow-angle lens position gives you a relatively (75) *shallow* (76) *wide* (77) *great* depth of field.

4. The figure below shows the camera zoomed in all the way for a telephoto view and focused on object A. Object B will probably be (78) *in focus* (79) *out of focus.* The depth of field is therefore (80) *great* (81) *shallow.*

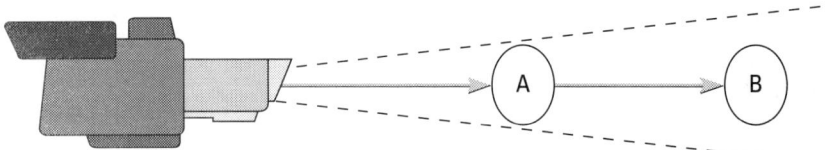

5. The figure below shows the camera zoomed out all the way for a wide-angle view and focused on object A. Object B will probably be (82) *in focus* (83) *out of focus.* The depth of field is therefore (84) *great* (85) *shallow.*

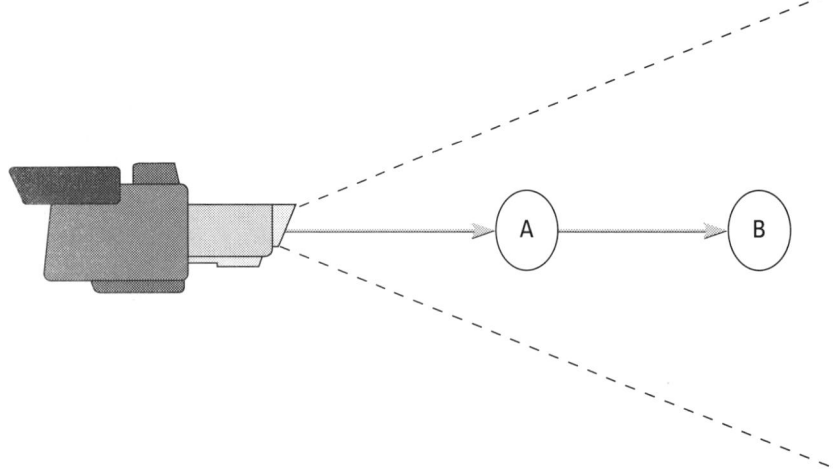

Part II — Image Creation and Control

Course No. _____ Date _____ Name _____

6. The screen image below displays a (86) *great* (87) *shallow* depth of field.

6 ○ 86 ○ 87

7. The screen image below shows that the camera's zoom lens was in a (88) *wide-angle* (89) *narrow-angle* position.

7 ○ 88 ○ 89

8. Your preview monitors for cameras 1, 2, and 3 display the following images. Assuming that all three cameras are positioned right next to one another, which is the approximate zoom position for each? Choose among (90) *wide angle (short focal length)* (91) *normal (medium focal length)* or (92) *narrow angle (long focal length)*.

a—Camera 1   b—Camera 2   c—Camera 3

8a ○ 90 ○ 91 ○ 92

8b ○ 90 ○ 91 ○ 92

8c ○ 90 ○ 91 ○ 92

PAGE TOTAL

SECTION TOTAL

Chapter 4 — Looking Through the Viewfinder

REVIEW QUIZ

Mark the following statements as true or false by filling in the bubbles in the T (for true) or F (for false) column.

1. Leadroom and noseroom fulfill similar framing (compositional) functions.
2. The aperture of a lens influences the depth of field.
3. Wide-angle zoom positions slow down perceived z-axis speed.
4. There is no aesthetic difference between a zoom and a dolly.
5. When following lateral motion, you should keep as little space as possible between the object and the screen edge toward which the object is moving.
6. Psychological closure always ensures good composition.
7. In general, video is more of a close-up than a long-shot medium.
8. Close-ups usually have a shallow depth of field.
9. A close-up should provide visual clues for closure in off-screen space.
10. You should always try to achieve psychological closure within the TV screen area.

Course No. _____ Date _____ Name _____

ZETTL'S VIDEO LAB 2.0 QUIZ

Click on the **camera** monitor and take the quizzes on tape 4 **Focusing**, tape 5 **Screen Forces**, tape 6 **Composition**, tape 7 **Picture Depth**, and tape 8 **Screen Motion**.

PROBLEM-SOLVING APPLICATIONS

1. Zoom all the way out with your camcorder, or attach a wide-angle lens (28mm or less focal length) to your 35mm camera, and focus on an object about 4 to 6 feet away from you. Look at the background objects (about 20 or so feet away from you). Are they visible? Do they appear in fairly sharp focus? Or are they blurred? Next, make the same observations by zooming all the way in or by attaching a telephoto lens (with a focal length of 200mm) to your still camera. Now relate your observations to depth of field.

2. When watching television or a movie, try to figure out what lenses were used for some of the shots. For example, when you see someone running toward the camera yet seemingly not getting closer, what lens was used? Or when you see the happy couple approach the dinner table through the out-of-focus flowers and candles in the foreground, what lens was probably used? Such observations will help you become more aware of focal lengths and their effects. Compare your notes with others watching the same program or movie.

3. The director tells you, the camera operator, to change from a cross-shot to an over-the-shoulder shot. What does the director mean? How can you accomplish such a shot change?

4. The director wants the foreground object in focus but the background out of focus. How can you accomplish this request?

5. Your documentary is to show the congestion and lack of breathing space between the units in a new suburban housing development. When aiming your camera along the street, what zoom position would you use? Why?

6. With your camcorder in the wide-angle zoom position, walk toward an object. Go back to your starting point and this time zoom in on the object. When playing back the two scenes, can you tell which was the dolly and which was the zoom? How?

Chapter 4 — Looking Through the Viewfinder

Course No. _____ Date _____ Name _____

 # Operating the Camera

REVIEW OF KEY TERMS

Match each term with its appropriate definition by filling in the corresponding bubble.

1. arc
2. dolly
3. cant
4. crane
5. truck
6. tongue
7. mounting head
8. tilt
9. pedestal
10. pan
11. white balance
12. calibrate zoom lens

A. To move the camera laterally by means of a mobile camera mount.

B. To move the camera in a slightly curved dolly or truck.

C. To move the boom with the camera from left to right or from right to left.

D. A device that connects the camera to its support.

PAGE TOTAL

Chapter 5 — Operating the Camera

1. arc	5. truck	9. pedestal
2. dolly	6. tongue	10. pan
3. cant	7. mounting head	11. white balance
4. crane	8. tilt	12. calibrate zoom lens

E. To point the camera up or down.

F. To move the camera toward or away from the object.

G. The adjustment of the color channels in the camera to produce a white color in lighting of various color temperatures.

H. To preset a zoom lens in order to keep in focus throughout the zoom.

I. Horizontal turning of the camera.

J. To move the camera up or down with a studio camera mount.

Course No. _____ Date _____ Name _____

1. arc	5. truck	9. pedestal
2. dolly	6. tongue	10. pan
3. cant	7. mounting head	11. white balance
4. crane	8. tilt	12. calibrate zoom lens

K. Tilting the camera sideways.

L. To move the boom of the camera crane up or down.

REVIEW OF CAMERA MOUNTS

Select the correct answers, and fill in the bubbles with the corresponding numbers.

1. A spreader (13) *keeps the tripod legs from spreading too far* (14) *helps to spread the tripod legs as much as possible* (15) *maximizes the spread of the tripod legs.*

2. The wedge mount (16) *facilitates the mounting of a camera on the mounting head* (17) *secures the dolly base to the tripod legs.*

Chapter 5 — Operating the Camera

3. Fill in the bubbles whose numbers correspond with the camera movement indicated in the following figure.

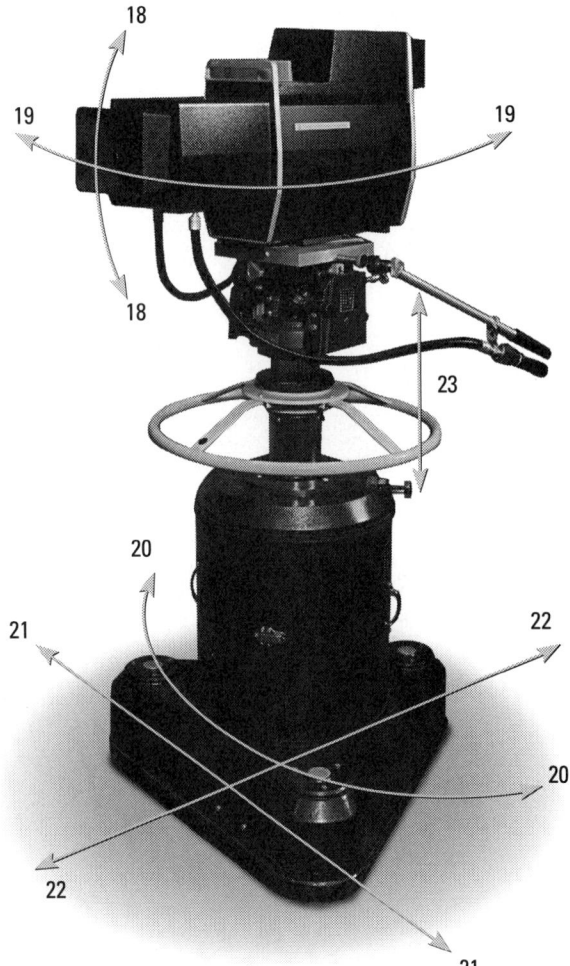

a. dolly

b. truck

c. tilt

3a ○ 18   ○ 19   ○ 20
    ○ 21   ○ 22   ○ 23

3b ○ 18   ○ 19   ○ 20
    ○ 21   ○ 22   ○ 23

3c ○ 18   ○ 19   ○ 20
    ○ 21   ○ 22   ○ 23

PAGE TOTAL

Course No. _____ Date _____ Name _____

d. pan

3d ○ ○ ○
   18 19 20
   ○ ○ ○
   21 22 23

e. pedestal

3e ○ ○ ○
   18 19 20
   ○ ○ ○
   21 22 23

f. arc

3f ○ ○ ○
   18 19 20
   ○ ○ ○
   21 22 23

4. Identify in the following figure whether the dolly wheels on this studio pedestal are set for (24) *parallel* (25) *tricycle* or (26) *freewheeling* steering.

4 ○ ○ ○
  24 25 26

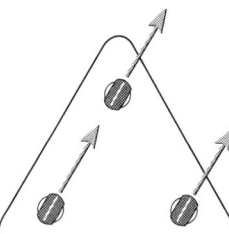

5. For all normal camera moves, the studio pedestal is normally set for (27) *parallel* (28) *tricycle* (29) *freewheeling* steering.

5 ○ ○ ○
  27 28 29

6. To minimize the wiggles of a handheld or shoulder-mounted camcorder, you should have your camera lens (30) *zoomed out all the way (wide-angle position)* (31) *zoomed in all the way (narrow-angle position)* (32) *zoomed halfway in (normal focal-length position)*.

6 ○ ○ ○
  30 31 32

PAGE TOTAL ☐
SECTION TOTAL ☐

Chapter 5 — Operating the Camera

REVIEW OF OPERATIONAL FEATURES

Select the correct answers, and fill in the bubbles with the corresponding numbers.

1. When white-balancing an ENG/EFP camera for covering a school board meeting in the school's multipurpose room, you should white-balance your camera (33) *in the hallway before entering the room* (34) *with a special white-balancing light* (35) *in the light that actually illuminates the area where the school board members sit.*

   1 ○ 33   ○ 34   ○ 35

2. Calibrating, or presetting, the zoom lens means (36) *adjusting the zoom lens so that it keeps in focus throughout the zoom* (37) *setting the proper f-stop* (38) *adjusting the zoom range.*

   2 ○ 36   ○ 37   ○ 38

3. Proper calibration of the zoom lens requires that you (39) *zoom in* (40) *zoom out,* and focus on the target object (41) *closest* (42) *farthest* from your camera.

   3 ○ 39   ○ 40
      ○ 41   ○ 42

4. The "memory" of a battery refers to (43) *an automatic indicator that tells when the battery was in use* (44) *an indicator that tells when the battery is empty* (45) *an indication of a full charge when the battery is only partially charged.*

   4 ○ 43   ○ 44   ○ 45

REVIEW QUIZ

*Mark the following statements as true or false by filling in the bubbles in the **T** (for true) or **F** (for false) column.*

1. The studio pedestal is an ideal remote camera mount.

2. The camera mounting heads are important only for studio cameras but not for ENG/EFP cameras.

3. The automatic focus on a camcorder guarantees keeping in focus at all times.

4. The wedge mount ensures that the camera is mounted in an optimally balanced position.

5. The jib arm and the camera crane can make the camera move in similar ways.

6. Dolly and truck movements show up as similar movements on the screen.

7. The professional ENG/EFP camera requires new white-balancing every time the camera is moved from one lighting environment to the next.

8. To boom up means to raise the camera pedestal.

9. The depth of field depends on the focal length of the lens, the lens aperture, and the distance from camera to object.

10. With a crane, you can boom, tongue, and pan the camera in one single motion.

11. You need to lock your camera mounting head only when leaving the camera unattended for a long period of time.

12. Calibrating your zoom lens is necessary for ENG/EFP cameras as well as for studio cameras.

13. Generally, tight close-ups have a shallow depth of field.

14. You can ignore the camcorder's "low battery" warning if you have recently charged this battery.

15. Zooming all the way out will minimize camera wobbles.

Chapter 5 — Operating the Camera

ZETTL'S VIDEO LAB 2.0 QUIZ

Click on the **camera** monitor and take the quizzes on tape 2 **Zoom Lens** and tape 9 **Camera Moves**; then click on the **lights** monitor and take the quiz on tape 5 **Color Temperature**.

PROBLEM-SOLVING APPLICATIONS

1. Locate the tilt and pan drag and lock mechanisms of your camera. Adjust them so that you can pan and tilt the camera as smoothly as necessary. When do you need to use the lock mechanisms?

2. Place three chairs along the z-axis about 9 feet apart. Zoom in to an ECU on the first chair. It will probably be out of focus at the end of your zoom. Now bring your picture into focus. When zooming out, the picture will remain in focus. Without touching your focus, now zoom in on the last chair. Again, your picture will probably get out of focus when reaching the ECU position. Now focus on the last chair and zoom back. Without touching your focus, zoom in on the first chair, zoom back, and then zoom in again on the last chair. Will the first as well as the last chair remain in focus during the zoom-in? If so, why? If not, why not?

3. With your knees pointing in the direction of the start of the pan, pan your camera slowly and smoothly about 180 degrees. Now repeat the same pan with your knees preset as much as possible in the direction of the end point of the pan. Which position makes for a smoother pan? Why?

4. With the automatic white balance turned off, videotape for a few seconds a white object (such as a small white card) outdoors, then repeat the same shot with indoor lighting (such as with your reading lamp shining on it), and then under fluorescent lights. Compare the various color tints the "white" card has taken on. Repeat the same procedure but with the automatic white balance turned on or by white-balancing for each of the illuminations. Now do the colors of the white object look the same in each shot? Why?

5. With a handheld or shoulder-mounted camcorder, follow a friend from behind and videotape him or her while walking along the z-axis. Now do the same while walking backward and having your friend face your camera during the videotaping. Which version made it easier for you to keep the camera steady? Why?

Course No. _____ Date _____ Name _____

6 Light, Color, Lighting

REVIEW OF KEY TERMS

Match each term with its appropriate definition by filling in the corresponding bubble.

1. directional light
2. diffused light
3. baselight
4. lux
5. foot-candle
6. attached shadow
7. cast shadow
8. falloff
9. additive primary colors
10. color temperature

A. Light that illuminates a relatively small area with a distinct light beam. Usually produced by spotlights, it creates harsh, clearly defined shadows.

B. Even, nondirectional light necessary for the camera to operate optimally. Refers to the overall light intensity.

C. The speed (degree) with which a light picture portion turns into shadow areas. It can be fast or slow.

D. The American unit for measuring light intensity.

E. Standard European unit for measuring light intensity.

Chapter 6 — Light, Color, Lighting

1. directional light
2. diffused light
3. baselight
4. lux
5. foot-candle
6. attached shadow
7. cast shadow
8. falloff
9. additive primary colors
10. color temperature

F. Light that illuminates a relatively large area with an indistinct light beam.

G. Red, green, and blue.

H. Shadow that is produced by an object and thrown on another surface. It can be independent of the object.

I. Shadow that is on the object itself. It cannot be seen independently of the object.

J. Relative reddishness or bluishness of light, as measured in Kelvin degrees.

Course No. _____ Date _____ Name _____

Match each term with its appropriate definition by filling in the corresponding bubble.

11. **light intensity**
12. **spotlight**
13. **floodlight**
14. **contrast**
15. **photographic principle**
16. **key light**
17. **fill light**
18. **back light**
19. **background light**
20. **light plot**

K. A plan, similar to a floor plan, that shows the type, size (wattage), and location of the lighting instruments relative to the scene to be illuminated and the general direction of the beams.

L. Additional light on the opposite side of the camera from the key light to illuminate shadow areas and thereby reduce falloff. Usually done with floodlights.

M. The difference between the lightest and darkest spot in a picture.

N. Illumination of the set pieces and backdrop.

O. The triangular arrangement of key, back, and fill lights, with the back light opposite the camera and directly behind the object, and the key and fill lights on opposite sides of the camera and to the front and side of the object.

P. Illumination from behind the subject and opposite the camera.

Q. Principal source of illumination. Usually a spotlight.

Chapter 6 — Light, Color, Lighting

11. light intensity	16. key light
12. spotlight	17. fill light
13. floodlight	18. back light
14. contrast	19. background light
15. photographic principle	20. light plot

R. Lighting instrument that produces diffused light.

S. Lighting instrument that produces directional, relatively undiffused light.

T. The amount of light reflected off an object seen by the lens.

Course No. _____ Date _____ Name _____

REVIEW OF LIGHT, SHADOWS, COLOR

Select the correct answers, and fill in the bubbles with the corresponding numbers.

1. When lighting for slow falloff in large areas, you should use predominantly (21) *directional* (22) *diffused* (23) *high-color-temperature* light.

2. Baselight levels are measured in (24) *lux* (25) *foot-candles* (26) *f-stops*. [Multiple answers are possible.]

3. One foot-candle is approximately (27) *5* (28) *10* (29) *15* lux.

4. Fast falloff means that there is (30) *a great difference between the light side and attached shadow side of the object* (31) *a great difference between the light side and cast shadow of the object* (32) *little difference between the light side and attached shadow side of the object*.

5. A high color temperature means that the light has a (33) *reddish* (34) *bluish* (35) *greenish* tinge.

6. The standard color temperature for indoor lighting is (36) *3,200°K* (37) *3,600°K* (38) *5,600°K*.

7. The standard color temperature for outdoor lighting is (39) *3,200°K* (40) *3,600°K* (41) *5,600°K*.

8. Electronic white balance (42) *makes a white object look white on the monitor, regardless of the relative color temperature of the light* (43) *controls extremely bright spots in the picture* (44) *adjusts the brightness in the viewfinder*.

9. When all three electron guns in a color monitor hit the three additive color dots at maximum intensity, the screen will be (45) *white* (46) *black* (47) *multicolored*.

10. When lowering the color temperature, the white light becomes more (48) *reddish* (49) *greenish* (50) *bluish*.

SECTION TOTAL

Chapter 6 — Light, Color, Lighting

REVIEW OF LIGHTING INSTRUMENTS

1. Fill in the bubbles whose numbers correspond with the appropriate lighting instruments shown below:

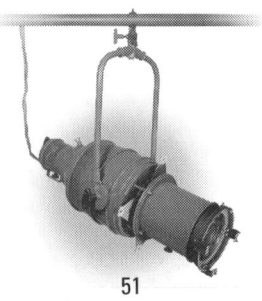
51

52

53

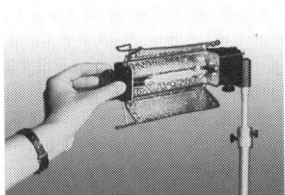
54

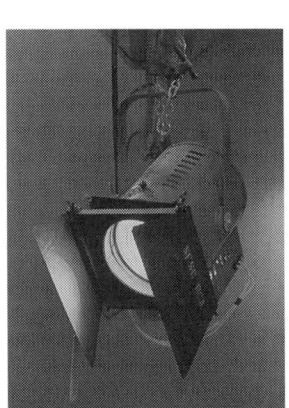
55

56

57

58

59

Part II — Image Creation and Control

Course No. _____ Date _____ Name _____

a. Omni light

1a ○ ○ ○ ○ ○
51 52 53 54 55
○ ○ ○ ○
56 57 58 59

b. strip, or cyc, light

1b ○ ○ ○ ○ ○
51 52 53 54 55
○ ○ ○ ○
56 57 58 59

c. EFP floodlight (Tota light)

1c ○ ○ ○ ○ ○
51 52 53 54 55
○ ○ ○ ○
56 57 58 59

d. Fresnel spotlight

1d ○ ○ ○ ○ ○
51 52 53 54 55
○ ○ ○ ○
56 57 58 59

e. scoop

1e ○ ○ ○ ○ ○
51 52 53 54 55
○ ○ ○ ○
56 57 58 59

f. ellipsoidal spotlight

1f ○ ○ ○ ○ ○
51 52 53 54 55
○ ○ ○ ○
56 57 58 59

g. softlight

1g ○ ○ ○ ○ ○
51 52 53 54 55
○ ○ ○ ○
56 57 58 59

h. clip light

1h ○ ○ ○ ○ ○
51 52 53 54 55
○ ○ ○ ○
56 57 58 59

i. large broad

1i ○ ○ ○ ○ ○
51 52 53 54 55
○ ○ ○ ○
56 57 58 59

PAGE TOTAL

Chapter 6 — Light, Color, Lighting

Select the correct answers, and fill in the bubbles with the corresponding numbers.

2. To flood (spread) the light beam of a Fresnel spotlight, you need to move the bulb-reflector assembly (60) *toward* (61) *away from* the lens; to focus the beam more narrowly, move it (62) *toward* (63) *away from* the lens.

3. The flooded beam has (64) *more* (65) *less* intensity than the focused beam.

4. Portable spotlights have (66) *a Fresnel zoom lens* (67) *no lens* and (68) *a small focus control* (69) *no focus control.*

5. A scoop has (70) *a Fresnel lens* (71) *no lens* and (72) *a focus control* (73) *no focus control.*

Part II — Image Creation and Control

Course No. _____ Date _____ Name _____

REVIEW OF LIGHTING TECHNIQUES

Select the correct answers, and fill in the bubbles with the corresponding numbers.

1. The arrangement of lighting instruments shown in the figure below is generally called (74) *the basic photographic principle* (75) *four-point lighting* (76) *photographic lighting* (77) *the field lighting principle.*

   1 ○ ○ ○ ○
      74 75 76 77

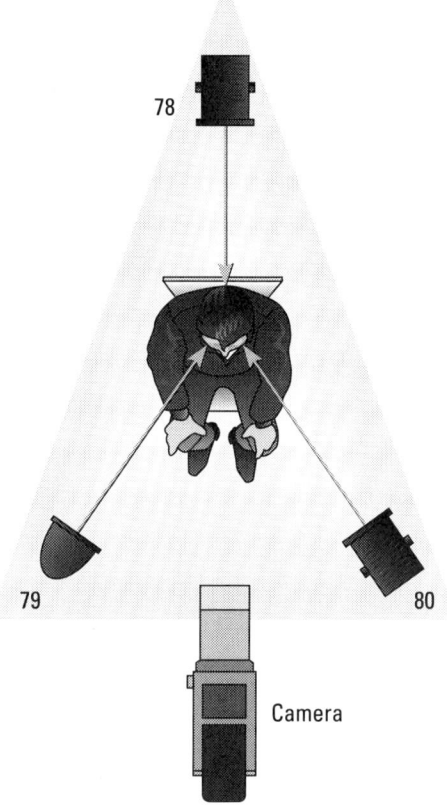

2. Fill in the bubbles whose numbers correspond with the functions of lighting instruments shown in the figure above and whether they are usually (81) *spotlights* or (82) *floodlights.*

   a. key

   2a ○   ○   ○
       78   79   80
          ○   ○
          81   82

   b. back

   2b ○   ○   ○
       78   79   80
          ○   ○
          81   82

   c. fill

   2c ○   ○   ○
       78   79   80
          ○   ○
          81   82

   PAGE TOTAL [ ]

Chapter 6 — *Light, Color, Lighting*

55

3. What major light sources were used for illuminating the talk-show host in the following three pictures? In the diagrams, circle the instrument or instruments used, and then fill in the bubbles whose numbers correspond with the instruments used to light the subject.

a.

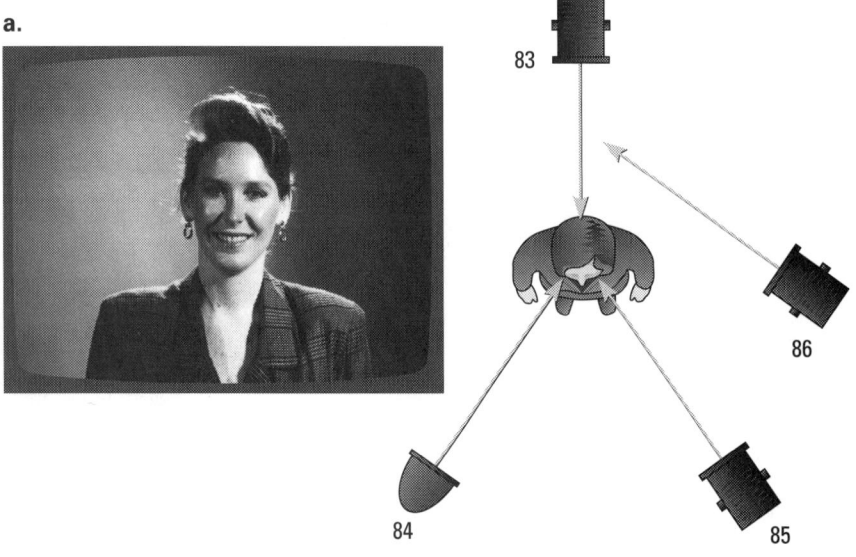
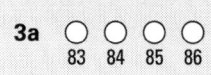

3a ○ ○ ○ ○
   83 84 85 86

b.

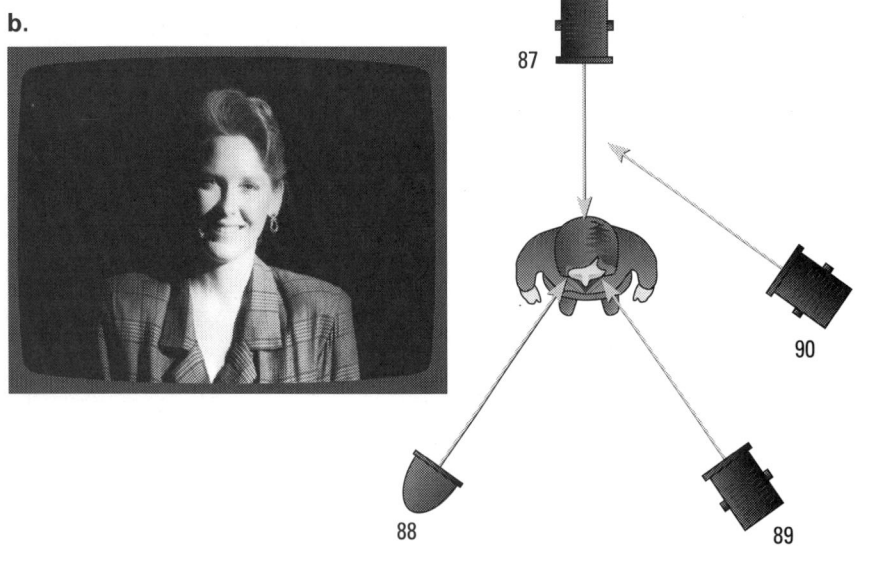
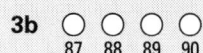

3b ○ ○ ○ ○
   87 88 89 90

PAGE TOTAL

56

Part II — *Image Creation and Control*

Course No. _____ Date _____ Name _____

c.

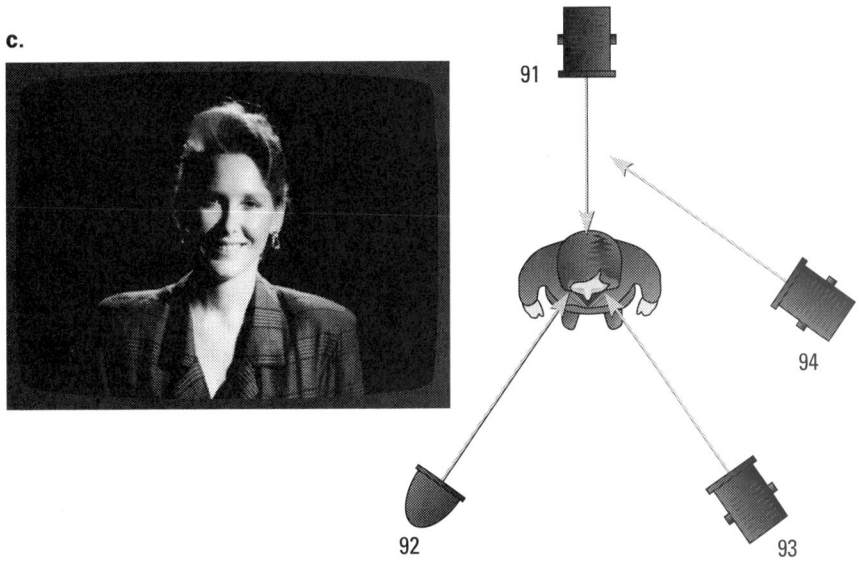

3c ○ ○ ○ ○
    91 92 93 94

4. When shooting an ENG interview in bright sunlight, the most convenient fill light is (95) *a large spot* (96) *a scoop* (97) *a reflector*.

4 ○   ○   ○
   95 96 97

5. Having somebody stand in front of a brightly illuminated building will (98) *provide much needed back light* (99) *help to separate the person from the background* (100) *cause an undesirable silhouette effect*.

5 ○   ○   ○
   98 99 100

6. When trying to match the color temperature of outdoor light coming into a room through a window with that of your standard portable floodlights, you need to (101) *lower* (102) *raise* the color temperature of your indoor lights by putting an (103) *amber (warm yellow)* (104) *light blue* gel (filter) in front of all indoor instruments.

6 ○   ○
   101 102
   ○   ○
   103 104

PAGE TOTAL _____
SECTION TOTAL _____

Chapter 6 — Light, Color, Lighting

57

REVIEW QUIZ

Mark the following statements as true or false by filling in the bubbles in the T (for true) or F (for false) column.

1. The basic photographic principle uses a key light, a fill light, and a back light.
2. The background light must strike the background from the same side as the key light.
3. A bluish gel can raise the color temperature.
4. The more fill light, the slower the falloff.
5. All professional floodlights have Fresnel lenses.
6. Color temperature refers to how hot a light burns.
7. A cast shadow that touches the base of the lighted object becomes an attached shadow.
8. Back lights and background lights fulfill similar functions.
9. Floodlights are the quickest way to illuminate a large area with even light.
10. The best lighting instrument for EFP is the 2,000-watt Fresnel spotlight.
11. We measure color temperature on the Kelvin scale.
12. Barn doors are primarily used to slow down falloff.
13. Small ENG/EFP lights, such as the Lowel Omni light, have no lens.
14. C-clamps are built for mounting spotlights but not floodlights.
15. Softlights can focus their light beam.

Course No. _____ Date _____ Name _____

ZETTL'S VIDEO LAB 2.0 QUIZ

Click on the **lights** monitor and take the quizzes on tape 2 **Light and Shadow**, tape 3 **Falloff**, tape 4 **Measurement**, tape 6 **Instruments**, tape 7 **Triangle Lighting**, tape 8 **Design**, and tape 9 **Field**.

PROBLEM-SOLVING APPLICATIONS

1. You are asked to do the lighting for a shampoo commercial. The director wants you to make the model's blond hair look especially brilliant and glamorous. Which of the three instruments of the lighting triangle needs special attention to achieve the desired result?

2. You have very little time to light a five-member panel discussion in the multipurpose room of the local elementary school. The board members sit side-by-side behind a long table. What lighting type would you employ? What instruments would you use? Why?

3. The director of a fashion show complains about the extremely fast falloff lighting and wants you to slow down the falloff for her spring fashion show. What does she mean? How can you accommodate her wishes?

4. You are asked to videotape the president of a new computer company who wants to talk to his employees from behind his desk. The vice president of the computer company tells you not to worry about the large window behind the president's chair because you can use the daylight streaming through the window as interesting back light. What is your reaction? What problems, if any, do you anticipate? What solutions do you suggest?

5. The novice assistant director is very much worried about videotaping the high-school graduation ceremony. He said that the overcast sky will cause fast falloff and is prone to color distortion in the shadow areas. What is your reaction? Why?

Chapter 6 — *Light, Color, Lighting*

Course No. _____ Date _____ Name _____

Visual Effects and Computer-Generated Video

REVIEW OF COMPUTER TERMINOLOGY

Match each term with its appropriate definition by filling in the corresponding bubble.

1. analog
2. baud
3. scanner
4. byte
5. compression
6. digital
7. digitize
8. hard drive
9. file
10. DVD

A. A CD-ROM–like high-capacity disc that can store up to 4.7 gigabytes of information.

B. A signal that fluctuates exactly like the original stimulus.

C. A device that translates visual images (such as this page) into digital information.

D. A specific collection of information stored on the disk separately from other information.

E. A series of 8 bits.

PAGE TOTAL

Chapter 7 — Visual Effects and Computer-Generated Video

1. analog
2. baud
3. scanner
4. byte
5. compression
6. digital
7. digitize
8. hard drive
9. file
10. DVD

F. A process by which the data are neatly ordered so that they do not waste any available disk space, or the data are reduced during the storage process and restored during the retrieval.

G. Pertaining to data in the form of digits (on/off pulses).

H. A high-capacity disk that is built into the computer or connected to the computer by cable.

I. Data transmission speed, expressed in signal events per second.

J. To convert an analog signal into digital (binary) form or to transfer information into a digital code.

Course No. _____ Date _____ Name _____

Match each term with its appropriate definition by filling in the corresponding bubble.

11. **gigabyte**
12. **initialize**
13. **megabyte**
14. **menu**
15. **menu bar**
16. **modem**
17. **pixel**
18. **platform**
19. **RAM**
20. **ROM**

K. To prepare a disk so that it can receive and store digital information in an orderly fashion.

L. Roughly 1,074 million bytes.

M. The strip across the top of the display screen that shows available options.

N. Short for picture element. The smallest picture element, like the dot in a newspaper picture. In computer graphics, the smallest visual unit that can be addressed and processed by the graphics program.

O. Stands for read-only memory, the program that is built into the computer memory and cannot be altered.

P. Stands for random-access memory. It is actually a read/write memory chip that makes possible storage and retrieval of information while the computer is in use.

Q. Stands for modulator/demodulator. Equipment that changes the digital computer signals into analog signals and back again.

Chapter 7 — *Visual Effects and Computer-Generated Video*

11. gigabyte
12. initialize
13. megabyte
14. menu
15. menu bar
16. modem
17. pixel
18. platform
19. RAM
20. ROM

R. Designates the specific operating system (Windows, Macintosh).

S. A list of the material stored or a set of options displayed after loading a computer program.

T. Roughly 1,048,000 bytes.

REVIEW OF THE DESKTOP COMPUTER SYSTEM

Select the correct answers, and fill in the bubbles with the corresponding numbers.

1. RAM (21) *loses* (22) *remembers* (23) *transfers* its information each time the computer is turned off.

2. ROM (24) *loses* (25) *remembers* (26) *transfers* its information each time the computer is turned off.

3. The storage capacity of a floppy disk is (27) *less than* (28) *the same as* (29) *more than* that of a hard drive.

4. The CD-ROM is a (30) *read/write* (31) *read-only* (32) *write-only* storage device.

5. The modem translates exclusively (33) *digital data into analog form* (34) *analog data into digital form* (35) *digital data into analog form and vice versa.*

6. The scanner translates (36) *alphanumeric characters but no pictures* (37) *pictures but no alphanumeric characters* (38) *both alphanumeric characters and pictures* into digital form.

Course No. _____ Date _____ Name _____

REVIEW OF KEY TERMS

Match each term with its appropriate definition by filling in the corresponding bubble.

39. ESS system
40. DVE
41. super
42. C.G.
43. key
44. wipe
45. chroma key
46. matte key

A. Analog electronic visual effect in which the base image replaces the black areas of the foreground picture.

B. Stores many still video frames in digital form for easy access.

C. Video effects generated by a computer.

D. The simultaneous overlay of two pictures on the same screen.

E. Special effect that uses color (usually blue) for the background of the source (foreground picture). All blue areas are replaced by the base picture.

F. Letters of a keyed title are filled with gray or a specific color through a third video source.

G. A small computer dedicated to the creation of letters and numbers.

H. A transition in which one image gradually replaces the other on the screen.

Chapter 7 — Visual Effects and Computer-Generated Video

REVIEW OF STANDARD ELECTRONIC EFFECTS

1. Fill in the bubbles whose numbers correspond with the appropriate key effects shown in the following figures.

47

48

49

50

    a. matte key   1a ○ ○ ○ ○
                                                           47 48 49 50

    b. outline mode   1b ○ ○ ○ ○
                                                           47 48 49 50

    c. drop-shadow mode   1c ○ ○ ○ ○
                                                           47 48 49 50

    d. edge mode   1d ○ ○ ○ ○
                                                           47 48 49 50

PAGE TOTAL

Course No. _____ Date _____ Name _____

2. Fill in the bubbles whose numbers correspond with the buttons you would have to press on the pattern selector to create the various effects (a through e) illustrated in the following figures.

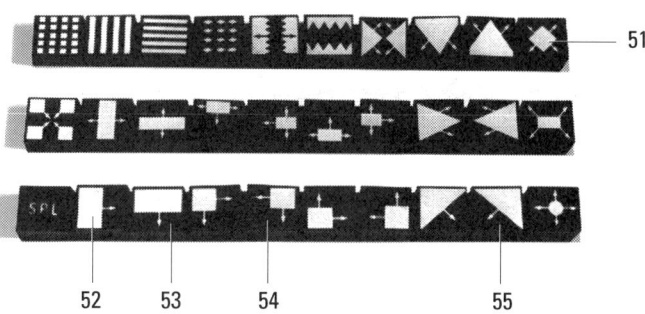

a.

b.

c.

2a ○ ○ ○ ○ ○
   51 52 53 54 55

2b ○ ○ ○ ○ ○
   51 52 53 54 55

2c ○ ○ ○ ○ ○
   51 52 53 54 55

PAGE TOTAL

Chapter 7 — *Visual Effects and Computer-Generated Video*

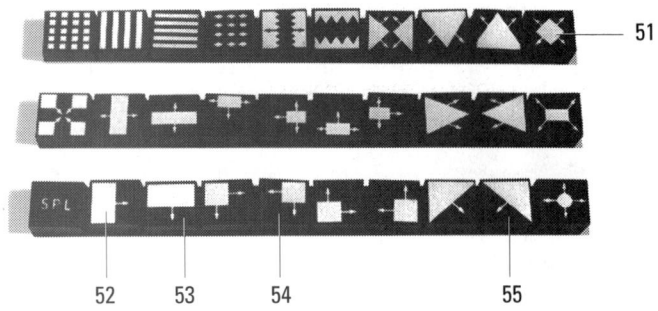

— 51

52 53 54   55

d.

e.

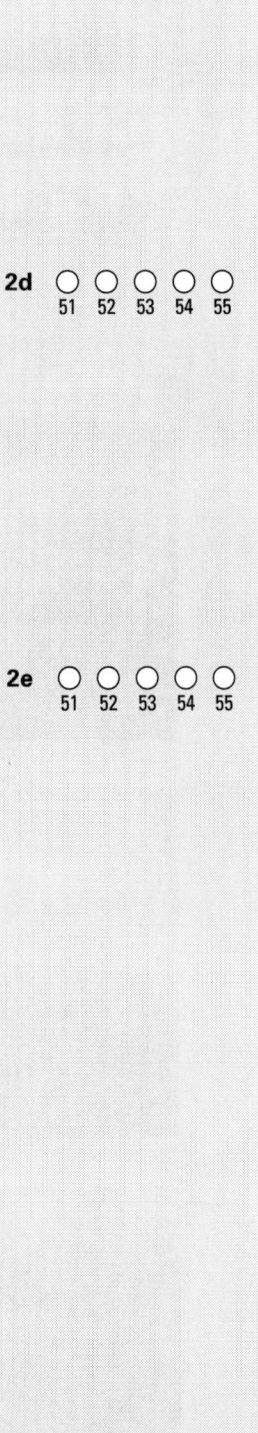

68

Part II — Image Creation and Control

Course No. _____ Date _____ Name _____

3. Fill in the bubbles whose numbers correspond with the appropriate electronic effects shown in the following figures.

56

57

58

59

60

a. cube effect

b. fly effect

c. mosaic effect

d. solarization

e. horizontal stretching

Chapter 7 — Visual Effects and Computer-Generated Video

REVIEW QUIZ

*Mark the following statements as true or false by filling in the bubbles in the **T** (for true) or **F** (for false) column.*

1. You can use a green background for chroma keying.

2. A superimposition has the foreground image block out portions of the background image as though the foreground image were cut into the background image.

3. You can achieve a split-screen effect simply by stopping a horizontal wipe in mid-center.

4. The ESS system functions like a large slide library and superfast slide projector.

5. When chroma keying, the object that is to be keyed into the background must be blue.

6. Before the C.G. can be used for a title, you need to print the title first on a studio card.

7. All pictures that appear on the video screen must be camera-generated.

8. Morphing is a digital effect that translates an analog image into a digital one.

9. Filling keyed letters with a specific color is called a matte key.

10. A CD-ROM is a read/write device.

11. RAM will lose its information when the computer power is switched off.

12. A 10-megabyte hard drive (disk) has a larger storage capacity than a 10-gigabyte one.

13. A scanner translates digital signals into analog ones.

14. The higher the baud rate of a modem, the faster the transmission of data.

15. All information stored in ROM will be lost when the computer is switched off.

	T	F
1	61	62
2	63	64
3	65	66
4	67	68
5	69	70
6	71	72
7	73	74
8	75	76
9	77	78
10	79	80
11	81	82
12	83	84
13	85	86
14	87	88
15	89	90

SECTION TOTAL

Course No. _____ Date _____ Name _____

ZETTL'S VIDEO LAB 2.0 QUIZ

Click on the **editing** monitor and take the quiz on tape 6 **Transitions & Keys**.

PROBLEM-SOLVING APPLICATIONS

1. Your AD informs you that a dancer, who is supposed to be chroma-keyed over a videotaped landscape scene, is wearing a saturated medium-blue leotard. Your TD is very much concerned about this news. But your PA assures you that the floor manager has already taken care of the problem. What was the potential problem? What could the floor manager do to solve this problem?

2. The preview monitor shows that the outline mode title key is hard to read over the busy background. How could you make the title more readable without changing either font or background?

3. You are the director of a new international interview show. Although the guests do not normally come to the studio personally but participate in the interview via satellite hookup, you nevertheless want each guest to appear on the same screen with the interviewer. Your TD tells you that this cannot be done because your switcher is not equipped with DVE equipment. What is your response?

4. A friend tells you that a 12,000-baud speed is preferable to a 28,800-baud speed because the slower speed has a higher information capacity and, therefore, takes less time for the actual transmission of data. Do you agree with your friend? Why or why not?

5. The same friend tells you that you will no longer have to worry about losing information through a power failure or computer malfunction because your new computer has a 32-megabyte RAM. What is your response?

Chapter 7 — *Visual Effects and Computer-Generated Video*

Course No. _____ Date _____ Name _____

Audio and Sound Control

REVIEW OF KEY TERMS

Match each term with its appropriate definition by filling in the corresponding bubble.

1. omnidirectional
2. pickup pattern
3. hypercardioid
4. polar pattern
5. cardioid
6. unidirectional
7. ribbon microphone
8. dynamic mic
9. windscreen
10. lavaliere

A. High-quality, highly sensitive microphone for critical sound pickup. Produces warm sound.

B. A small microphone that is clipped onto clothing.

C. A highly directional microphone that can make faraway sounds appear relatively close.

D. Acoustic foam rubber that is put over the microphone to cut down wind noise in outdoor use.

E. The territory around the microphone within which the mic can hear well.

Chapter 8 — Audio and Sound Control

1. omnidirectional
2. pickup pattern
3. hypercardioid
4. polar pattern
5. cardioid
6. unidirectional
7. ribbon microphone
8. dynamic mic
9. windscreen
10. lavaliere

F. A unidirectional, heart-shaped microphone pickup pattern.

G. The microphone can hear best from the front.

H. The microphone can hear equally well from all directions.

I. The two-dimensional representation of the pickup pattern.

J. A relatively rugged microphone. Good for outdoor use.

Course No. _____ Date _____ Name _____

Match each term with its appropriate definition by filling in the corresponding bubble.

11. VU meter
12. jack
13. DAT
14. ATR
15. XLR
16. sweetening
17. mini plug
18. RCA phono plug
19. condenser microphone
20. fader

K. A professional three-wire connector for audio cables.

L. Stands for audiotape recorder.

M. Small connector normally used for most consumer video and audio equipment.

N. A sliding volume control.

O. Measures volume units, the relative loudness of amplified sound.

P. Tiny connector used for some consumer audio equipment.

Q. Stands for digital audiotape.

Chapter 8 — Audio and Sound Control

75

11. VU meter
12. jack
13. DAT
14. ATR
15. XLR
16. sweetening
17. mini plug
18. RCA phono plug
19. condenser microphone
20. fader

R. High-quality microphone with a battery power supply.

S. A socket or receptacle for a connector.

T. Manipulating recorded sound in postproduction.

Course No. _____ Date _____ Name _____

REVIEW OF SOUND-GENERATING ELEMENTS AND SOUND PICKUP

Select the correct answers, and fill in the bubbles with the corresponding numbers.

1. Select the three types of microphones as classified by how they are made—their sound-generating element: (21) *cardioid* (22) *ribbon* (23) *omnidirectional* (24) *dynamic* (25) *condenser* (26) *unidirectional*.

2. Fill in the bubbles whose numbers correspond with the polar patterns in the following figure.

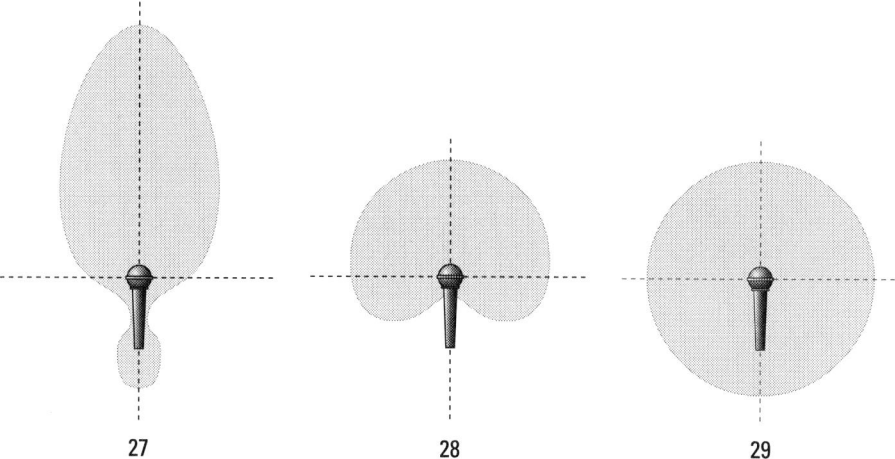

27   28   29

   a. omnidirectional

   b. cardioid

   c. hypercardioid

3. A shotgun mic has a pickup pattern that is (30) *omnidirectional* (31) *unidirectional (hypercardioid)* (32) *nondirectional*.

4. The most rugged microphones that work well outdoors are (33) *dynamic* (34) *ribbon* (35) *condenser*.

5. To eliminate sudden breath pops when speaking close to the microphone, use a (36) *pop filter* (37) *windscreen* (38) *breath filter*.

6. The microphone that needs a special power supply (usually batteries) to amplify the sound signal in the microphone is (39) *dynamic* (40) *ribbon* (41) *condenser*.

Chapter 8 — *Audio and Sound Control*

REVIEW OF MICROPHONE USE

Select the correct answers, and fill in the bubbles with the corresponding numbers.

1. Boom microphones (fishpole and perambulator) normally have a(n) (42) *cardioid* (43) *omnidirectional* (44) *hyper- or supercardioid* pickup pattern.

2. The hand microphones used in ENG normally have a(n) (45) *omnidirectional* (46) *bidirectional* (47) *hyper- or supercardioid* pickup pattern.

3. The most rugged ENG hand mics have (48) *dynamic* (49) *ribbon* (50) *condenser* sound-generating elements.

4. Mark *two* of the high-quality types of hand microphones normally used by singers: (51) *dynamic* (52) *ribbon* (53) *condenser.*

5. You are to set up microphones for a six-member panel discussion. All participants sit in a row at a special panel table. Normally, you would use (54) *boom mics* (55) *hand mics* (56) *desk mics* for this production.

	42	43	44
1	○	○	○

	45	46	47
2	○	○	○

	48	49	50
3	○	○	○

	51	52	53
4	○	○	○

	54	55	56
5	○	○	○

SECTION TOTAL

Part II — Image Creation and Control

Course No. _____ Date _____ Name _____

REVIEW OF SOUND CONTROL

Select the correct answers, and fill in the bubbles with the corresponding numbers.

1. Overloading the volume of the incoming sound signal will result in (57) *distorted sound* (58) *earache* (59) *damage to the VU meter*.

2. The variety of quality controls on an audio console is (60) *greater than* (61) *the same as* (62) *less than* the quality controls on an audio mixer.

3. The test tone at the beginning of a videotape recording should be (63) *–1VU* (64) *+1VU* (65) *0 VU*.

Fill in the bubble whose number corresponds with the number identifying the equipment shown in the following two figures.

4. The appropriate place that marks the beginning of the "overload" zone is (66) *–5VU* (67) *–2VU* (68) *0 VU*.

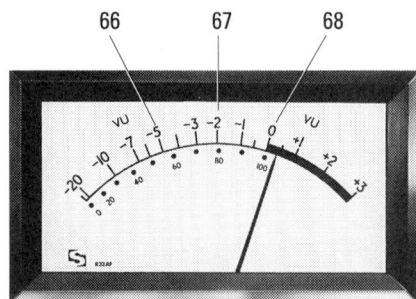

5. The various heads in the head assembly shown in the following figure are:

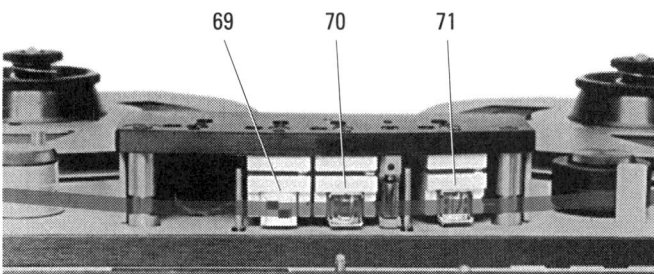

   a. playback head

   b. erase head

   c. record head

Chapter 8 — Audio and Sound Control

REVIEW OF SOUND RECORDING AND AESTHETICS

Select the correct answers, and fill in the bubbles with the corresponding numbers.

1. A 24-track recorder needs the following number of head assemblies (erase, record, and playback): (72) *12* (73) *24* (74) *48*.

2. A CD contains audio information in (75) *laser* (76) *analog* (77) *digital* form.

3. The figure-ground principle in audio refers to (78) *a person speaking while moving against a stable background* (79) *making the principal sound source as loud as the background sounds* (80) *separating the principal sound source from the background sounds through a higher volume.*

4. Matching audio and video energies means to (81) *adjust the strength of audio signals to that of video signals* (82) *use special digital video and audio devices* (83) *have high- or low-energy audio accompany high- or low-energy video.*

Course No. _____ Date _____ Name _____

REVIEW QUIZ

Mark the following statements as true or false by filling in the bubbles in the T (for true) or F (for false) column.

		T	F

1. Lavaliere microphones can be used for some types of music pickup.

2. Dynamic mics are generally less sensitive to shock and temperature extremes than ribbon mics.

3. Because lavaliere microphones are highly sensitive, they work best when hidden under a shirt or blouse.

4. To test whether a microphone is turned on, you should blow into it.

5. Regardless of whether you work with a mixer or audio console, each input has its own pot (fader).

6. Riding gain means adjusting the VU meter so that it reads 0 regardless of the input audio levels.

7. In EFP you should try to mix the sounds as carefully as possible with the use of a portable mixer.

8. DAT cassettes cannot be played on regular (analog) cassette players.

9. All sounds used on a videotape must first be captured by a microphone and recorded on audiotape.

10. All microphones are insensitive to shock once they are turned off.

SECTION TOTAL

Chapter 8 — Audio and Sound Control

ZETTL'S VIDEO LAB 2.0 QUIZ

Click on the **audio** monitor and take the quizzes on tape 3 **Microphones**, tape 4 **Connectors**, tape 5 **Mixers**, and tape 6 **Aesthetics**.

PROBLEM-SOLVING APPLICATIONS

1. You are in charge of audio for a show that consists of several intimate numbers by a singer and a small band. After the rehearsal, an observer in the control room tells you that the singer holds the mic during the softer passages much too close to her mouth, and that she should hold the mic much lower and sing "across" rather than into it. What is your reaction? Why?

2. You are doing a documentary on police patrols in your city. You first want to hear the conversations and the police radio inside the patrol car, and then capture the sounds of possible conversations, yelling, or any other audio when the officers leave the patrol car to confront a suspect. What microphones would you need for optimal sound pickup in these situations?

3. While riding gain during a live-on-tape pickup of a small rock band, the director is concerned about overmodulation when the VU meters occasionally peak into the +1 red zone. What is your response?

4. During a small segment of an EFP in an auto assembly plant, the new Triple-I intern wonders why the audio people do not try mixing the ambient sounds with the voices of the reporter and the plant supervisor right on the spot. He feels that mixing in the field would eliminate a great deal of postproduction sweetening. What are your comments?

5. The same intern tells you that the audio engineer of his former employer used lavaliere microphones exclusively on the actors of a college play because these mics would provide excellent sound perspective without driving the audio console operator crazy. Do you agree with this audio engineer? If so, why? If not, why not?

Part III

Video Recording, Storage, and Sequencing

Course No. _____ Date _____ Name _____

Video Recording

REVIEW OF KEY TERMS

Match each term with its appropriate definition by filling in the corresponding bubble.

1. audio track
2. video track
3. analog VTR system
4. Y/C component system
5. composite system
6. multimedia
7. RGB component system
8. interactive video
9. nonlinear storage system
10. time base corrector (TBC)
11. digital VTR
12. control track

A. The area of the videotape used for recording the audio information.

B. The area of the videotape used for recording the synchronization information.

C. The area of the videotape used for recording the video information.

D. A videotape system that records the video signal in analog form.

Chapter 9 — Video Recording

85

1. audio track
2. video track
3. analog VTR system
4. Y/C component system
5. composite system
6. multimedia
7. RGB component system
8. interactive video
9. nonlinear storage system
10. time base corrector (TBC)
11. digital VTR
12. control track

E. A videotape system that records the video signal in digital form.

F. A system that combines the Y (black-and-white) and C (red, green, and blue) video information into a single signal.

G. A system in which all three color signals are kept separate throughout the video recording process.

H. Separating the Y and C signals and treating the luminance and color as separate signals.

I. An electronic accessory to videotape recorders that helps to make playbacks or transfers electronically stable. It keeps slightly different scanning cycles in step.

J. The simultaneous computer display of text, still and moving images, and sound. Usually recorded on CD-ROMs.

Course No. _____ Date _____ Name _____

---

1. audio track
2. video track
3. analog VTR system
4. Y/C component system
5. composite system
6. multimedia
7. RGB component system
8. interactive video
9. nonlinear storage system
10. time base corrector (TBC)
11. digital VTR
12. control track

K. Storage of video and audio material in digital form on a hard drive or read/write optical disc. Each frame can be accessed by the computer independently of all others.

K ○ ○ ○ ○
    1 2 3 4
    ○ ○ ○ ○
    5 6 7 8
    ○ ○ ○ ○
    9 10 11 12

L. A computer-driven video program that gives the viewer some control over what to see and how to see it. It is often used as a training device.

L ○ ○ ○ ○
    1 2 3 4
    ○ ○ ○ ○
    5 6 7 8
    ○ ○ ○ ○
    9 10 11 12

PAGE TOTAL [ ]

SECTION TOTAL [ ]

Chapter 9 — *Video Recording*

REVIEW OF VIDEOTAPE RECORDING SYSTEMS

Select the correct answers, and fill in the bubbles with the corresponding numbers.

1. The NTSC signal is (13) *composite* (14) *component* and needs (15) *one wire* (16) *two wires* (17) *three wires* to be transported.

2. The RGB signal is (18) *composite* (19) *component* and needs (20) *one wire* (21) *two wires* (22) *three wires* to be transported.

3. The Y/C component system needs (23) *one* (24) *two* (25) *three* wires to be transported.

4. In order to keep the quality loss to a minimum during extensive post-production editing, you should use (26) *analog* (27) *digital* VTRs and a (28) *Y/C* (29) *RGB* (30) *NTSC* system.

5. All ½-inch VTR formats are (31) *compatible* (32) *not compatible* and, therefore, (33) *can* (34) *cannot* all be played back on a regular VHS videotape machine.

6. You (35) *can* (36) *cannot* play a regular VHS tape on an S-VHS recorder, but you (37) *can* (38) *cannot* play an S-VHS tape on a regular VHS recorder.

7. D-2 and DVCPRO systems are (39) *compatible* (40) *not compatible*, which means that you (41) *can* (42) *cannot* play back a DVCPRO tape on a D-2 machine and vice versa.

8. Because DVCPRO and DVCAM are (43) *digital* (44) *analog* systems, their signals (45) *need not* (46) *need to* be digitized for nonlinear editing.

SECTION TOTAL

Part III — Video Recording, Storage, and Sequencing

Course No. _____ Date _____ Name _____

REVIEW OF THE VIDEO RECORDING PROCESS

Select the correct answers, and fill in the bubbles with the corresponding numbers.

1. In order to record on a videocassette, the protection tab must be in the (47) *open* (48) *closed* position, or (49) *broken off* (50) *unbroken*.

2. Mark the items that are part of a normal video leader: (51) *color bars* (52) *identification slate* (53) *control track display* (54) *test tone* (55) *edit decision list* (56) *video test pattern*. **(Multiple answers are possible.)**

3. Color bars are useful only when they are (57) *generated by the equipment you are actually using* (58) *dubbed from a color bar master tape* (59) *digitally generated.*

1 ◯ ◯
   47 48
   ◯ ◯
   49 50

2 ◯ ◯ ◯
   51 52 53
   ◯ ◯ ◯
   54 55 56

3 ◯ ◯ ◯
   57 59 59

SECTION TOTAL [   ]

REVIEW OF NONLINEAR STORAGE SYSTEMS

Select the correct answers, and fill in the bubbles with the corresponding numbers.

1. In the following list, the "read-only" digital storage device is a (60) *floppy disk* (61) *hard drive* (62) *digital VTR* (63) *CD-ROM*.

2. Indicate which of the following digital storage devices are nonlinear: (64) *DVD* (65) *hard drive* (66) *digital VTR* (67) *ESS*. **(Multiple answers are possible.)**

3. The ESS system can grab and digitize a video frame (68) *only from a digital videotape* (69) *only from an analog videotape* (70) *from any video source.*

1 ◯ ◯ ◯ ◯
   60 61 62 63

2 ◯ ◯ ◯ ◯
   64 65 66 67

3 ◯ ◯ ◯
   68 69 70

SECTION TOTAL [   ]

Chapter 9 — Video Recording

REVIEW OF INTERACTIVE VIDEO AND MULTIMEDIA

Select the correct answers, and fill in the bubbles with the corresponding numbers.

1. Interactive video means (71) *the opportunity for the viewer to participate in the production of the program* (72) *the opportunity for the viewer to exercise some choice in the presented material with immediate feedback* (73) *the opportunity for the viewer to interact with the producer of the program.*

2. Multimedia commonly refers to the simultaneous computer display of (74) *text and still images, but no moving images* (75) *text, and still and moving images, but no sound* (76) *text, still and moving images, and sound.*

3. A multimedia program can be used only with (77) *a computer* (78) *a VTR* (79) *a VTR coupled with a computer.*

Course No. _____ Date _____ Name _____

REVIEW QUIZ

*Mark the following statements as true or false by filling in the bubbles in the **T** (for true) or **F** (for false) column.*

1. In an RGB component system, the three primary color signals are kept separate throughout the recording process.

2. The Y/C component system means that the color yellow has been added to the color signals.

3. NTSC signal and composite signal mean the same thing.

4. Color bars help in adjusting colors of the playback monitor.

5. A video leader should always be dubbed over from a master recording of standard video leaders.

6. A TBC helps eliminate picture jitter.

7. You can play back a BetacamSP videotape on an S-VHS recorder.

8. You can play back a VHS videotape on an S-VHS recorder.

9. The difference between D-2 and DVCAM systems is that the D-2 system is digital, whereas the DVCAM system is analog.

10. You can replace your old VCR recorder with a D-2 VTR without changing any of the other production components, such as monitors and switchers.

11. The video leader includes a 0 VU test tone.

12. Dubbing analog videotape produces more deterioration from one generation to the next than digital videotape.

13. In order to record on a cassette, the tab must be intact or in the closed position.

14. You can record a new program on a CD-ROM, but you will lose the old program in the process.

15. Multimedia programs cannot display moving images.

SECTION TOTAL

Chapter 9 — Video Recording

ZETTL'S VIDEO LAB 2.0 QUIZ

Click on the **editing** monitor and take the quiz on tape 3 **Tape Basics**.

PROBLEM-SOLVING APPLICATIONS

1. Your assistant shows you the first draft of an entry form for a statewide video competition. The specifications for tape formats read as follows: "Only ½-inch or Hi8 tape formats will be accepted." Will you recommend any changes to this entry? If so, why? If not, why not?

2. The budget manager of a corporate production company tells you that she would immediately buy a D-2 VTR if it would not require the replacement of all other existing analog equipment. What is your reply? Why?

3. You are in a meeting in which one of the award-winning old-timers in television production tells you that "the smaller the tape format, the less video quality you can expect." Do you agree with this statement? If so, why? If not, why not?

4. Your director tells you that you need to dub your S-VHS master to a regular VHS format in order to play it on your regular home VCR. Do you agree? If so, why? If not, why not?

5. A new intern tells you that you can substitute a digital VTR for an ESS system and call up the stored video images just as fast. What are your comments?

Course No. _____ Date _____ Name _____

10 Editing Principles

REVIEW OF KEY TERMS

Match each term with its appropriate definition by filling in the corresponding bubble.

1. continuity editing
2. complexity editing
3. continuing vectors
4. mental map
5. diverging vectors
6. jump cut
7. converging vectors
8. vector line
9. jogging
10. cutaway

A. A shot that is used to intercut between two shots in which the screen direction is reversed.

B. Frame-by-frame advancement of videotape, resulting in jerky motion.

C. The assembly of shots to ensure vector continuity.

D. Graphic vectors that extend each other, or index and motion vectors pointing and moving in the same direction.

E. The building of an intensified screen event from carefully selected and juxtaposed shots.

Chapter 10 — Editing Principles

1. continuity editing
2. complexity editing
3. continuing vectors
4. mental map
5. diverging vectors
6. jump cut
7. converging vectors
8. vector line
9. jogging
10. cutaway

F. Tells us where things are or are supposed to be on- and off-screen.

G. Index and motion vectors that point toward each other.

H. Index and motion vectors that point away from each other.

I. An image that jumps slightly from one screen position to another during a cut.

J. An imaginary line created by extending converging index vectors, or the direction of a motion vector.

Course No. _____ Date _____ Name _____

REVIEW OF AESTHETIC PRINCIPLES OF CONTINUITY EDITING

Select the correct answers, and fill in the bubbles with the corresponding numbers.

1. In continuity editing, you are primarily concerned with maintaining (11) *story continuity* (12) *schedule efficiency* (13) *vector continuity.*

2. The mental map includes (14) *only on-screen positions and directions* (15) *only off-screen positions and directions* (16) *both on- and off-screen positions and directions.*

3. In order to preserve graphic vector continuity, you (17) *cannot cross the vector line* (18) *cross the vector line* (19) *ignore the vector line.*

4. In order to preserve index vector continuity, you (20) *must cross the vector line* (21) *should not cross the vector line* (22) *ignore the vector line.*

5. In order to preserve motion vector continuity, you (23) *must cross the vector line* (24) *should not cross the vector line* (25) *ignore the vector line.*

6. Vectors indicate (26) *a direction* (27) *an edit command* (28) *a specific address code.*

7. A jump cut means that the subject is (29) *jumping up and down* (30) *causing the videotape to break up at the edit point* (31) *perceived as abruptly changing from one screen position to the next.*

8. In the following figures, select the camera that is in the *wrong* place for continuity editing, and fill in the corresponding bubble.

32

A

B

33

34

a. Over-the-shoulder shots of person A and person B.

Chapter 10 — Editing Principles

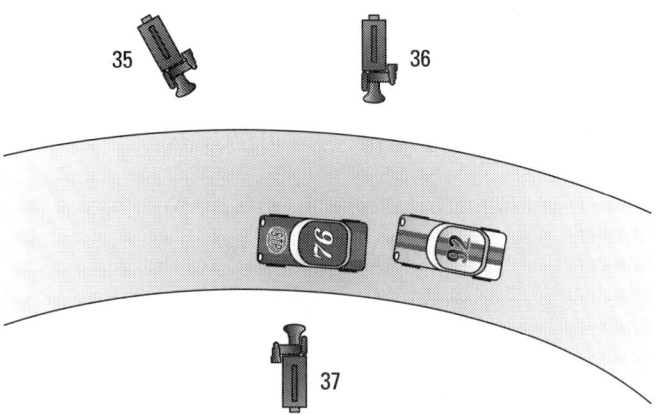

b. Cutting from long shots to close-ups during car race.

c. Cutting from speaker to audience.

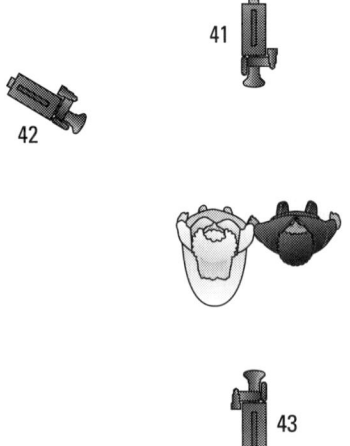

d. Cutting from camera 1 (41) to different points of view of the bride and groom during a wedding.

Course No. _____ Date _____ Name _____

9. For each of the following storyboard pairs, indicate whether the shots (44) *can* (45) *cannot* be edited together to form converging index vectors.

a.

b.

c.

d.

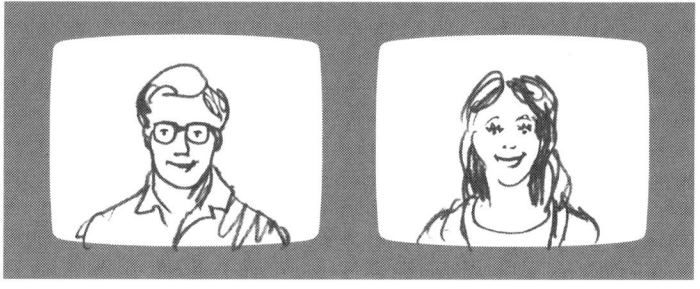

9a ○ ○
   44   45

9b ○ ○
   44   45

9c ○ ○
   44   45

9d ○ ○
   44   45

PAGE TOTAL ☐

Chapter 10 — *Editing Principles*

10. Cutting together the two shots shown below will result in (46) *a jump cut* (47) *diverging vectors* (48) *reversal of screen direction.*

Course No. _____ Date _____ Name _____

REVIEW QUIZ

*Mark the following statements as true or false by filling in the bubbles in the **T** (for true) or **F** (for false) column.*

1. Postproduction editing is done primarily to fix production mistakes.

2. Complexity editing requires a much stricter observance of the vector line principle than continuity editing.

3. The vector line is especially relevant for preserving screen positions and index and motion continuity.

4. A motion vector can be continuing or converging, but not diverging.

5. Z-axis index vectors can be converging or diverging, depending on context.

6. Motion vectors cannot converge in a single shot.

7. The mental map includes on-screen as well as off-screen positions.

8. When crossing the vector line during over-the-shoulder shooting, the talent will flip screen positions.

9. In complexity editing, a jump cut can be used to intensify the event.

10. The primary function of complexity editing is to correct production mistakes.

Chapter 10 — Editing Principles

ZETTL'S VIDEO LAB 2.0 QUIZ

*Click on the **editing** monitor and take the quizzes on tape 2 **Functions**, tape 4 **Continuity**, tape 5 **Location Procedures**, tape 7 **Cutting Procedures**, and tape 8 **Pre-edit Procedures**.*

PROBLEM-SOLVING APPLICATIONS

1. The new Triple-I intern suggests that you cover the New Year's parade by placing cameras exactly opposite each other on both sides of the street. Do you agree with his suggestion?

2. The host mispronounces the name of the guest during the videotaping of an interview. The director opts for going on rather than doing the opening again because "we can fix it in post" (postproduction). Do you agree with the director's decision? If so, why? If not, why not?

3. The director of a cable company production group tells you not to worry about what happens in off-screen space when editing, because the "viewer" will not see what is going on in off-screen space anyway. What is your response?

4. A retired, highly experienced film director tells you to avoid jump cuts at all costs. What is your reaction?

5. When you want to cross the vector line with your cameras for intensifying a scene, you are told by the producer that this is an absolute "no-no." How would you justify "crossing the line"?

Course No. _____ Date _____ Name _____

 Switching and Postproduction Editing

REVIEW OF KEY TERMS

Match each term with its appropriate definition by filling in the corresponding bubble.

1. program bus
2. mix buses
3. M/E bus
4. preview/preset bus
5. fader bar
6. downstream keyer
7. linear editing system
8. edit controller
9. nonlinear editing system
10. SMPTE time code
11. pulse-count system
12. off-line editing
13. window dub
14. on-line editing
15. EDL

A. It consists of edit-in and edit-out cues, expressed in time code numbers, and the nature and transitions between shots.

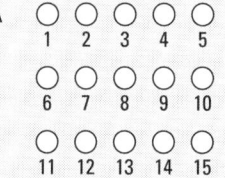

B. The bus (row of buttons) on the switcher, with inputs that are directly switched to the line-out.

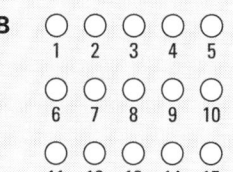

C. Produces the final high-quality edit master tape.

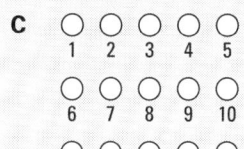

D. Rows of buttons that permit a super.

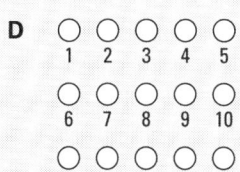

1. program bus
2. mix buses
3. M/E bus
4. preview/preset bus
5. fader bar
6. downstream keyer
7. linear editing system
8. edit controller
9. nonlinear editing system
10. SMPTE time code
11. pulse-count system
12. off-line editing
13. window dub
14. on-line editing
15. EDL

E. Refers to an editing process that will not produce an edit master tape, but a "rough-cut" or simply an edit decision list.

F. A lever on the switcher that produces transitions and effects of different speeds.

G. A control that allows captions to be keyed over the picture (line-out signal) as it leaves the switcher.

H. A single bus that can be assigned to perform either dissolves or wipes (with another M/E bus).

I. A device that facilitates various editing functions, such as marking edit-in and edit-out points.

J. Rows of buttons that can direct an input to the preview/preset monitor.

Course No. _____ Date _____ Name _____

1. program bus	6. downstream keyer	11. pulse-count system
2. mix buses	7. linear editing system	12. off-line editing
3. M/E bus	8. edit controller	13. window dub
4. preview/preset bus	9. nonlinear editing system	14. on-line editing
5. fader bar	10. SMPTE time code	15. EDL

K. Uses videotape as the recording medium. It does not allow random access of shots.

L. A specially generated address code that marks each video frame with a specific number.

M. An address code that uses the control track pulses to count elapsed time and frame numbers.

N. Allows random access of shots. The video and audio information is stored in digital form on computer disks.

O. A dub of the source tapes to a lower-quality tape format with the address code keyed into each frame.

Chapter 11— Switching and Postproduction Editing

REVIEW OF BASIC SWITCHER OPERATION

1. Fill in the bubbles whose numbers correspond with the appropriate parts of the switcher shown in the following figure.

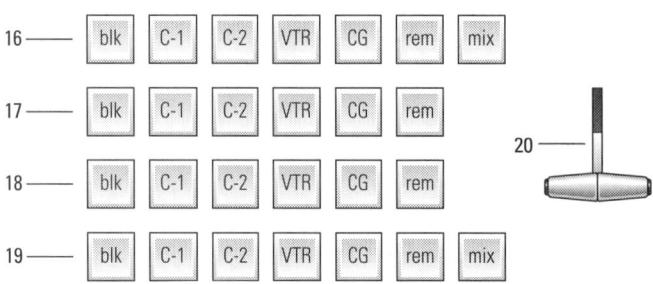

   a. mix bus A

   b. mix bus B

   c. preview bus

   d. program bus

   e. fader bar

2. Fill in the bubbles whose numbers correspond with the appropriate parts of the switcher shown in the following figure.

1a ○ ○ ○ ○ ○
   16 17 18 19 20

1b ○ ○ ○ ○ ○
   16 17 18 19 20

1c ○ ○ ○ ○ ○
   16 17 18 19 20

1d ○ ○ ○ ○ ○
   16 17 18 19 20

1e ○ ○ ○ ○ ○
   16 17 18 19 20

PAGE TOTAL

Part III — Video Recording, Storage, and Sequencing

Course No. _____ Date _____ Name _____

a. program bus

b. preset bus

c. fader bar

d. key bus

e. delegation controls (effects transition)

f. wipe pattern selector

g. downstream keyer controls

h. key/matte controls

Chapter 11 — *Switching and Postproduction Editing*

Select the correct answers, and fill in the bubbles with the corresponding numbers.

3. The program bus will direct the selected video source to the (29) *preview monitor* (30) *mix bus* (31) *line-out.*

4. Assuming that camera 1 is on the air, you can cut to VTR by pressing the (32) *key button* (33) *cut button* (34) *VTR button* on the program bus.

5. To select the functions of a specific bus or buses, you need to activate the (35) *downstream keyer* (36) *wipe pattern selector* (37) *specific delegation controls.*

6. Select the switching sequence that would produce a dissolve from camera 1 to VTR (assuming you are working with the Grass Valley 100 switcher on page 104 and that the background and mix delegation buttons are already activated): (38) *press C-1 on the program bus, press VTR on the preset bus, and move fader bar to the opposite position* (39) *press C-1 on the preset bus, press VTR on the program bus, and move fader bar to the opposite position* (40) *press C-1 on the program bus, press VTR on the key bus, and press the auto transition button.*

7. The highlighted buttons on the switcher below (Grass Valley 100) have already been pressed. In order to achieve a horizontal wipe, you still need to (41) *press the cut button* (42) *press the key delegation control* (43) *move the fader bar to the opposite position or press the auto transition button.*

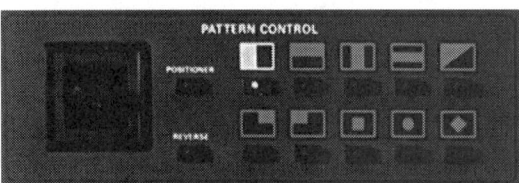

Course No. _____ Date _____ Name _____

REVIEW OF POSTPRODUCTION EDITING

Select the correct answers, and fill in the bubbles with the corresponding numbers.

1. Linear systems (44) *allow* (45) *do not allow* random access to the source material, and use (46) *videotape* (47) *digital storage devices* for their source material.

    1. ◯ 44   ◯ 45
       ◯ 46   ◯ 47

2. When using a normal linear single-source editing system, you can perform (48) *cuts and dissolves* (49) *cuts and wipes* (50) *wipes only* (51) *cuts only*.

    2. ◯ 48 ◯ 49 ◯ 50 ◯ 51

3. With a nonlinear system, you can create on the computer display screen (52) *cuts only* (53) *cuts and dissolves only* (54) *cuts, dissolves, and wipes* (55) *wipes only*.

    3. ◯ 52 ◯ 53 ◯ 54 ◯ 55

4. Select the function that is *not* performed by the edit controller: (56) *preroll the VTRs* (57) *simultaneously start play and record VTR* (58) *perform the actual edit* (59) *make the record VTR edit in either the assemble or insert mode*.

    4. ◯ 56 ◯ 57 ◯ 58 ◯ 59

5. The pulse-count editing system is (60) *more accurate* (61) *less accurate* in locating a specific frame than the time code system because it (62) *marks* (63) *does not mark* each individual frame with a specific address.

    5. ◯ 60   ◯ 61
       ◯ 62   ◯ 63

6. Because in the assemble editing mode the record VTR (64) *will* (65) *will not* erase the control track of the source tape, you (66) *need* (67) *do not need* to prerecord a continuous control track on the edit master tape.

    6. ◯ 64   ◯ 65
       ◯ 66   ◯ 67

7. Because in the insert editing mode the record VTR (68) *will* (69) *will not* transfer the control track from the source tape, you (70) *need* (71) *do not need* to prerecord a continuous control track on the edit master tape.

    7. ◯ 68   ◯ 69
       ◯ 70   ◯ 71

8. Fill in the bubbles whose numbers correspond with the numbers identifying the EDL mistakes in the following figure:

    8. ◯ 72   ◯ 73   ◯ 74
       ◯ 75   ◯ 76   ◯ 77

IN	OUT	
01 : 17 : 28 : 29	01 : 17 : 28 : 45	72
01 : 17 : 30 : 15	01 : 17 : 45 : 01	73
01 : 19 : 15 : 29	01 : 19 : 14 : 07	74
01 : 21 : 65 : 33	01 : 32 : 59 : 29	75
02 : 17 : 28 : 15	02 : 18 : 37 : 06	76
02 : 17 : 29 : 16	01 : 18 : 45 : 29	77

PAGE TOTAL [ ]

Chapter 11— *Switching and Postproduction Editing*

9. When editing in the insert mode, you need to record a continuous control track on (78) *all source tapes* (79) *the edit master tape* because the source VTRs (80) *transfer* (81) *do not transfer* their control track to the edit master tape.

10. When editing in the assemble mode, the source tapes (82) *will* (83) *will not* transfer their control tracks to the edit master tape, which (84) *requires* (85) *does not require* the recording of black on the edit master tape prior to editing.

Course No. _____ Date _____ Name _____

REVIEW QUIZ

Mark the following statements as true or false by filling in the bubbles in the T (for true) or F (for false) column.

1. The program bus on a switcher sends the selected source directly to the line-out.

2. Time code must be recorded with the actual production to achieve a continuous frame address.

3. Nonlinear editing equipment does not use source VTRs.

4. The edit controller for a linear editing system facilitates random access of various shots.

5. The auto transition button on a switcher fulfills the same function as the fader bar.

6. You should use insert editing only when replacing a shot in an edit master tape.

7. The pulse-count system marks each frame with a unique address.

8. The switcher has a separate button for each input.

9. Everything punched up on the key bus will go directly to the line-out.

10. On the Grass Valley 100 switcher, you can assign the key bus a mix function.

11. An EDL is necessary only if you do nonlinear editing.

12. The SMPTE time code marks each frame with a unique address.

13. A window dub keys a specific time code address over each frame.

14. Ordinarily, off-line editing does not produce the final edit master tape.

15. Nonlinear editing systems allow random access to pictures and sound.

Chapter 11— Switching and Postproduction Editing

ZETTL'S VIDEO LAB 2.0 QUIZ

Click on the **editing** monitor and take the quiz on tape 3 **Tape Basics**.

PROBLEM-SOLVING APPLICATIONS

1. Your director asks you, the TD, to superimpose a long shot of a dancer over the close-up of her face. First, the director wants to have the long shot as the more prominent image, and then slowly shift the emphasis in the super to the close-up of the dancer's face. How, if at all, can you accomplish such an effect?

2. Your director would like to have the total scene reveal itself gradually as though we were looking through a progressively expanding rectangle. She asks you to have the small rectangle start in the center of the screen with the proper TV aspect ratio and then expand out to the full-size television screen. How, if at all, could you achieve this effect?

3. You are asked by the corporate manager to use a nonlinear editing system for the daily company news show in order to save postproduction time. What is your response?

4. The same corporate manager tells you that he has denied your request for an edit controller because you have only a basic single-source editing system. However, he might be persuaded to buy one if you make a good argument for it. What is your argument?

5. You are told to skip the off-line editing step because your nonlinear desktop editing system will produce a high-quality edit master tape anyway. What is your comment?

Part IV

Talent and the Production Environment

Course No. _____ Date _____ Name _____

12 Talent, Clothing, and Makeup

REVIEW OF KEY TERMS

Match each term with its appropriate definition by filling in the corresponding bubble.

1. I.F.B.
2. moiré effect
3. actor
4. performer
5. talent
6. teleprompter
7. blocking
8. foundation
9. cue card

A. Collective name for all performers and actors who appear regularly on television.

B. A prompting system that allows communication with talent while on the air.

C. A device that projects the moving copy over the lens so that the talent can read it without losing eye contact with the viewer.

D. A large hand-lettered card that contains copy, usually held next to the camera lens by floor personnel.

E. Carefully worked-out movement and actions of the talent and all mobile video equipment used in a scene.

1. I.F.B.
2. moiré effect
3. actor
4. performer
5. talent
6. teleprompter
7. blocking
8. foundation
9. cue card

F. Color vibrations that occur when narrow, contrasting stripes of a design interfere with the scanning lines of the video system.

G. A person who appears on-camera in nondramatic shows.

H. A makeup base, normally done with water-soluble pancake makeup.

I. A person who appears on-camera in dramatic roles.

Course No. _____ Date _____ Name _____

REVIEW OF PERFORMING TECHNIQUES

1. The following figures show various cues as given to the talent by the floor manager. From the list below, select the specific cue illustrated, and fill in the bubble with the corresponding number.

   (10) *standby*   (15) *wind up*
   (11) *cue*   (16) *cut*
   (12) *on time*   (17) *4 minutes left*
   (13) *speed up*   (18) *½ minute left*
   (14) *stretch*   (19) *15 seconds left*

a. Pulls hands apart

b.

c.

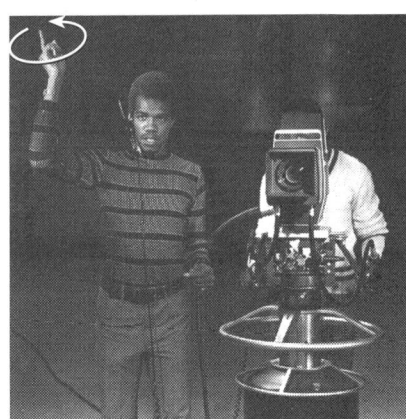

d. Rotates hand

1a ○ ○ ○ ○ ○
    10 11 12 13 14
    ○ ○ ○ ○ ○
    15 16 17 18 19

1b ○ ○ ○ ○ ○
    10 11 12 13 14
    ○ ○ ○ ○ ○
    15 16 17 18 19

1c ○ ○ ○ ○ ○
    10 11 12 13 14
    ○ ○ ○ ○ ○
    15 16 17 18 19

1d ○ ○ ○ ○ ○
    10 11 12 13 14
    ○ ○ ○ ○ ○
    15 16 17 18 19

PAGE TOTAL [ ]

Chapter 12 — *Talent, Clothing, and Makeup*

(10) *standby*	(15) *wind up*
(11) *cue*	(16) *cut*
(12) *on time*	(17) *4 minutes left*
(13) *speed up*	(18) *½ minute left*
(14) *stretch*	(19) *15 seconds left*

e.

f.

g. Rotates hand clockwise

h.

i.

j.

Part IV — Talent and the Production Environment

Course No. _____ Date _____ Name _____

2. The following figures show various cues as given to the talent by the floor manager. From the list below, select the specific cue illustrated, and fill in the bubble with the corresponding number.

(20) *VTR rolling*
(21) *closer*
(22) *back*
(23) *walk*
(24) *stop*

(25) *OK*
(26) *speak up*
(27) *tone down*
(28) *closer to mic*
(29) *keep talking*

a. Fingers open and close like a bird beak

b.

c. Pushes palms forward

d.

2a ◯ ◯ ◯ ◯ ◯
    20 21 22 23 24
    ◯ ◯ ◯ ◯ ◯
    25 26 27 28 29

2b ◯ ◯ ◯ ◯ ◯
    20 21 22 23 24
    ◯ ◯ ◯ ◯ ◯
    25 26 27 28 29

2c ◯ ◯ ◯ ◯ ◯
    20 21 22 23 24
    ◯ ◯ ◯ ◯ ◯
    25 26 27 28 29

2d ◯ ◯ ◯ ◯ ◯
    20 21 22 23 24
    ◯ ◯ ◯ ◯ ◯
    25 26 27 28 29

PAGE TOTAL _____

Chapter 12 — *Talent, Clothing, and Makeup*

(20) *VTR rolling*
(21) *closer*
(22) *back*
(23) *walk*
(24) *stop*

(25) *OK*
(26) *speak up*
(27) *tone down*
(28) *closer to mic*
(29) *keep talking*

e.

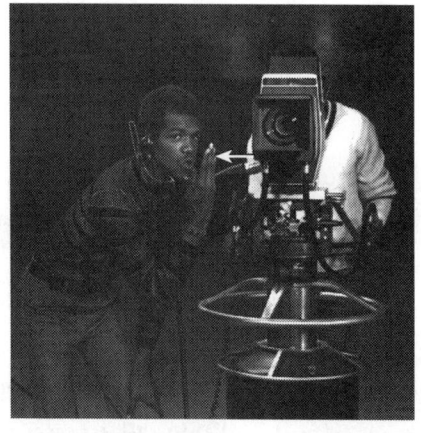
f.

g.

h. Pulls hands toward body

i. Moves fingers back and forth

j.

Course No. _____ Date _____ Name _____

Select the correct answers, and fill in the bubbles with the corresponding numbers.

3. For the talent the most accurate indicator of the camera's field of view is (30) *the relative distance between talent and camera* (31) *the floor manager's cues* (32) *the studio monitor.*

   3 ○ 30   ○ 31   ○ 32

4. If you do not use I.F.B. during a studio production, you should take your opening cues from (33) *the camera operator* (34) *the floor manager* (35) *the tally lights.*

   4 ○ 33   ○ 34   ○ 35

5. When you receive cues during the actual videotaping that are different from the rehearsed ones, you should (36) *execute the action as rehearsed* (37) *promptly follow the floor manager's cues* (38) *check first with the director.*

   5 ○ 36   ○ 37   ○ 38

6. When demonstrating a small object, you should (39) *hold it as close to the lens as possible* (40) *keep it as steady as possible or on the display table* (41) *move it for optimal camera pickup.*

   6 ○ 39   ○ 40   ○ 41

7. When asked for an audio level, you should (42) *blow into the mic* (43) *count quickly to three* (44) *recite your opening remarks at on-air levels.*

   7 ○ 42   ○ 43   ○ 44

8. When wearing a lavaliere mic, you should (45) *maintain your audio level regardless of how far the camera is away from you* (46) *increase your volume when the camera gets farther away from you* (47) *speak more softly when the camera is relatively close to you.*

   8 ○ 45   ○ 46   ○ 47

9. From the list below, select the microphone most appropriate for the various performance and acting tasks, and fill in the bubbles with the corresponding numbers.

   (48) *lavaliere mic*   (51) *stand mic*
   (49) *hand mic*   (52) *desk mic*
   (50) *fishpole mic*

   a. Interview with a celebrity at a busy airport gate.

   9a ○ 48 ○ 49 ○ 50 ○ 51 ○ 52

   b. News anchor who remains seated throughout newscast.

   9b ○ 48 ○ 49 ○ 50 ○ 51 ○ 52

   c. Moderating a panel discussion with six guests.

   9c ○ 48 ○ 49 ○ 50 ○ 51 ○ 52

   d. Lead guitarist of a rock band who also sings and talks to the audience.

   9d ○ 48 ○ 49 ○ 50 ○ 51 ○ 52

   e. Two actors doing a brief outdoor scene.

   9e ○ 48 ○ 49 ○ 50 ○ 51 ○ 52

   PAGE TOTAL ☐
   SECTION TOTAL ☐

Chapter 12 — *Talent, Clothing, and Makeup*

10. Read the following copy into a mirror or, better, into a television camera with a teleprompter, at least three times. Videotape your performances. Time your narration with a stopwatch and try to match your times with the ones given in subsequent readings.

a. news copy

VIDEO	AUDIO
TALENT ON CAMERA	IN SOUTHERN AUSTRALIA, ENGINEERS ARE RAVING ABOUT A NEW DEVELOPMENT...A GIANT EARTH MOVER THAT HAS NO WHEELS. CHIEF ENGINEER FRED STEINER SAYS HE OBSERVED HOW A CENTIPEDE TRAVELS AND SIMPLY COPIED ITS MOVEMENTS
VTR 6 V/O	THE MANY LEGS ENABLE THE MACHINE TO NEGOTIATE DITCHES, BUSHES, AND EVEN GOOD-SIZED BOULDERS WITHOUT SPILLING ITS LOAD OR TIPPING OVER.
VTR 4 VOT	JOHN HEWITT TALKED TO THE DRIVER...OR RIDER?...OF THE MONSTER CENTIPEDE...

Given time: 29 seconds

Your first reading: _____ seconds

Your second reading: _____ seconds

Your third reading: _____ seconds

Course No. _____ Date _____ Name _____

b. introduction to a weekly sports show

```
VIDEO                  AUDIO

STANDARD               Hi, I'm Alex Walter. Welcome
OPENING                to Sports at Four. Today we have
(VTR-SOT)              with us the world's most prominent
LIVE ON WALTER         and amazing


CU MESSNER             mountain climber--Reinhold Messner.
                       He will talk to us about how he
                       prepares for his extreme climbs, what
SLIDE: EVEREST         he thinks, and what he feels when he
                       climbs--often alone--the big walls of
                       the world's highest peaks.


MR WALTER              He has brought with him some of
                       the best mountain-climbing footage
                       I have ever seen--and he will share
                       it with us. We will also see where
                       Reinhold lives and what he does for
                       relaxation. All coming up next, on
                       Sports at Four.

------------------------------------------------------------
VTR 4                  COMMERCIAL #1
------------------------------------------------------------
```

Given time: 35 seconds

Your first reading: _____ seconds

Your second reading: _____ seconds

Your third reading: _____ seconds

Chapter 12 — *Talent, Clothing, and Makeup*

c. commercial
(Note that this commercial represents a high-pressure pitch and requires fast reading.)

VIDEO	AUDIO
OPEN CU OF BAKER ZOOM OUT TO MS	Hi, I'm Tom Baker of Baker Dodge to talk to you about automobile leasing. Now, leasing isn't only for big corporations; it's also for folks just like you. It can mean lower monthly payments, and it lets you keep that down payment for other things you need.
SUPER: LOWER MONTHLY PAYMENTS	
CUT TO: CAR 1	For example, you can lease this brand-new beauty for only 185 dollars per month, or this air-conditioned luxury car for only 250 dollars per month--and we'll even buy your present car and give you the cash!
CUT TO: CAR 2	
CUT TO: CU OF BAKER	So, come out now and see how our exclusive profit/lease plan can save you money!
CUT TO: SIGNATURE	It's only at Baker Dodge in Dodge City. Come right now and save!

Given time: 30 seconds

Your first reading: _____ seconds

Your second reading: _____ seconds

Your third reading: _____ seconds

Course No. _____ Date _____ Name _____

REVIEW OF ACTING TECHNIQUES

Select the correct answers, and fill in the bubbles with the corresponding numbers.

1. When doing an O/S or cross-shooting scene, you must adjust your blocking so that you see (53) *the key light* (54) *the floor manager* (55) *the camera lens.*

   1 ◯ 53 ◯ 54 ◯ 55

2. Pretend that you (person A) are receiving a telephone call from person B. In this scene we see and hear only A (you), but not B. Using exactly the same dialogue (see the script on the following page), adapt your delivery and acting style to at least two of the following circumstances:

   a. B calls to tell you that she is pregnant and unhappy about it.

   b. B is happy about being pregnant.

   c. B has just wrecked your new car.

   d. B has called off the wedding.

   e. B has won big in the lottery.

   f. B has been arrested.

   g. B has lost her job.

   h. B has won an Emmy for innovative productions.

   Place the scene anywhere you like. You may do well to write the other part of the phone conversation so that you can listen and respond more convincingly.

SECTION TOTAL

Chapter 12 — Talent, Clothing, and Makeup

123

PHONE CONVERSATION

Hello?
Hi.
Fine, and you?
Good.
No.
No, really. It's always a good time when you call.
I beg your pardon?
You must be kidding.
Yes.
No.
What does Alex say to all this?
No. Should I?
I don't know.
Perhaps.
You want me to come over now?
Yes. Really.
Well, this changes things somewhat.
I think so.
I'm not so sure.
Yes. No. I...
All right. But not...
OK.
If you think this is...
Definitely.
Good-bye...When?
No. Really.
Good-bye.

Course No. _____ Date _____ Name _____

REVIEW OF CLOTHING AND MAKEUP

Select the correct answers, and fill in the bubbles with the corresponding numbers.

1. Clothing with thin, highly contrasting stripes or checkered patterns is (56) *acceptable* (57) *not acceptable* because (58) *the CCD camera can handle such a contrast quite easily* (59) *it provides exciting patterns* (60) *it results in moiré color vibrations* (61) *it is too detailed for the camera to see.*

2. The dress of a pop singer has many rhinestones that sparkle under the colored stage lights. This dress is (62) *acceptable* (63) *unacceptable* because (64) *the color camera can handle small areas of extreme bright light* (65) *there is too much brightness contrast* (66) *it will result in moiré patterns* (67) *it will reinforce the stage lighting.*

3. During a normal chroma key, you should not wear (68) *red* (69) *blue* (70) *yellow* because this color will let the background show through during the keying.

4. One of the most widely used makeup foundations is (71) *pancake* (72) *grease base* (73) *oil-based foundation.*

5. When doing makeup, you should have lighting conditions that are the same as or close to those of (74) *your customary makeup room* (75) *the actual production environment* (76) *normal 3,200°K studio lights.*

1	○ 56	○ 57		
	○ 58	○ 59	○ 60	○ 61
2	○ 62	○ 63		
	○ 64	○ 65	○ 66	○ 67
3	○ 68	○ 69	○ 70	
4	○ 71	○ 72	○ 73	
5	○ 74	○ 75	○ 76	

SECTION TOTAL [   ]

Chapter 12 — Talent, Clothing, and Makeup

REVIEW QUIZ

*Mark the following statements as true or false by filling in the bubbles in the **T** (for true) or **F** (for false) column.*

1. Talent refers to both actors and performers.

2. You can ignore the time cues by the floor manager so long as you can see the studio clock.

3. In contrast to television performers, actors always portray someone else.

4. What you wear is really unimportant when simply auditioning for a role.

5. When working with a teleprompter, it is best to move the camera as close as possible to the talent.

6. A "cut" cue from the floor manager means that the director is cutting (switching) to the next camera.

7. Because you never know what you will be asked to read in an audition, you should not prepare for it but can and should rely on the energy of the moment during the audition.

8. Because of possible moiré effects, you should avoid wearing high-contrast striped patterns.

9. When on a close-up, you need to keep your actions limited and slower than normal.

10. You should try to get the floor manager's attention when you think that you should have received a time cue.

Course No. _____ Date _____ Name _____

ZETTL'S VIDEO LAB 2.0 QUIZ

There are no quizzes on the ZVL CD-ROM that ask specific questions about performing and acting. Watch and evaluate the performance and acting of the people you see on the disc.

PROBLEM-SOLVING APPLICATIONS

1. The talent for the new weekly *Opera Review* show arrives in a black tuxedo and a starched white dress shirt. The video operator is somewhat concerned about the talent's attire. Why? What would you suggest?

2. Your ENG camera operator is concerned about proper fill light for you, the reporter, during a live transmission of a traffic report on the fog-shrouded Golden Gate Bridge and asks for an additional person to handle the fill light. Do you share the camera operator's concern? If so, why? If not, why not?

3. The ENG camera operator suggests that you do your stand-up report right in front of the bright, sunlit wall of City Hall. According to the camera operator, the automatic iris control would guarantee the high-key lighting effect which, in turn, would reflect the upbeat story you have to tell. Do you agree with the camera operator? If so, why? If not, why not?

4. To practice blocking, write down a series of moves that carry you around your kitchen or living room. For example, you can start at the stove, then get the teakettle out of the cupboard, put it on the stove, go back to pick up the telephone, put down the telephone to answer the door, and so forth. Try to hit the same marks each time you go through the routine. If possible, have a friend videotape your blocking maneuvers from the same camera position. You can then compare the tapes and check how accurate your blocking was. As part of the same exercise, you can use various props (kitchen utensils) and see how the camera's field of view (LS to ECU) will influence your handling of props.

5. Do your favorite monologue and videotape it (or, better, have someone else tape it) first in a loose medium shot, then in a CU, and finally in an ECU. Analyze how the camera's field of view changes your delivery, and try to adjust your acting to each circumstance. Note especially the relative speed of your delivery and your facial expressions.

Chapter 12 — *Talent, Clothing, and Makeup*

Course No. _____ Date _____ Name _____

13 Production Environment: The Studio

REVIEW OF KEY TERMS

Match each term with its appropriate definition by filling in the corresponding bubble.

1. cyclorama
2. P.L.
3. S.A.
4. intercom
5. studio control room
6. monitor
7. master control
8. flats
9. props

A. Pieces of standing scenery used as background or to simulate the walls of a room.

B. Furniture and other objects used by talent and for set decoration.

C. A U-shaped continuous piece of canvas or muslin for backing of scenery and action.

D. High-quality video receiver used in the video studio and control rooms. Cannot receive broadcast signals.

E. Major intercommunication device in video studios.

Chapter 13 — Production Environment: The Studio

1. cyclorama
2. P.L.
3. S.A.
4. intercom
5. studio control room
6. monitor
7. master control
8. flats
9. props

F. Communication system for all production and engineering personnel involved in the production of a show. The most widely used system has telephone headsets to facilitate voice communication on several wired or wireless channels. Includes other systems, such as I.F.B. and cellular telephones.

G. A room adjacent to the studio in which the director, producer, various production assistants, the TD (technical director), the audio engineer, and sometimes the LD (lighting director) perform their various production functions.

H. Controls the program input, storage, and retrieval of on-the-air telecasts. Also oversees technical quality of all program material.

I. A public-address loudspeaker system from the control room to the studio.

130 *Part IV* — *Talent and the Production Environment*

Course No. _____ Date _____ Name _____

REVIEW OF VIDEO PRODUCTION STUDIO

Select the correct answers, and fill in the bubbles with the corresponding numbers.

1. Even for small studios, the minimum ceiling height is (10) *12 feet* (11) *14 feet* (12) *18 feet.*

2. Which of the following items is *not* part of the major studio installation: (13) *lights* (14) *camera and mic outlets* (15) *VTRs* (16) *intercom system* (17) *studio monitor* (18) *S.A. system.*

3. Assuming that the off-the-air receiver (AIR) shows the correct image, which preview monitors display the wrong picture?

   (19) *VT-1 monitor*   (26) *line monitor*
   (20) *VT-2 monitor*   (27) *C-1 monitor*
   (21) *VT-3 monitor*   (28) *C-2 monitor*
   (22) *character generator monitor*   (29) *C-3 monitor*
   (23) *electronic still store monitor* (30) *remote-1 monitor*
   (24) *effects monitor*   (31) *remote-2 monitor*
   (25) *preview monitor*

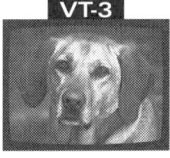

4. The switcher should be located adjacent to the (32) *LD's position* (33) *director's position* (34) *producer's position* (35) *C.G. operator's position.*

Chapter 13 — Production Environment: The Studio

REVIEW OF SCENERY, PROPERTIES, AND SCENIC DESIGN

1. Fill in the bubbles whose numbers correspond with the numbers identifying the various set pieces in the following figure.

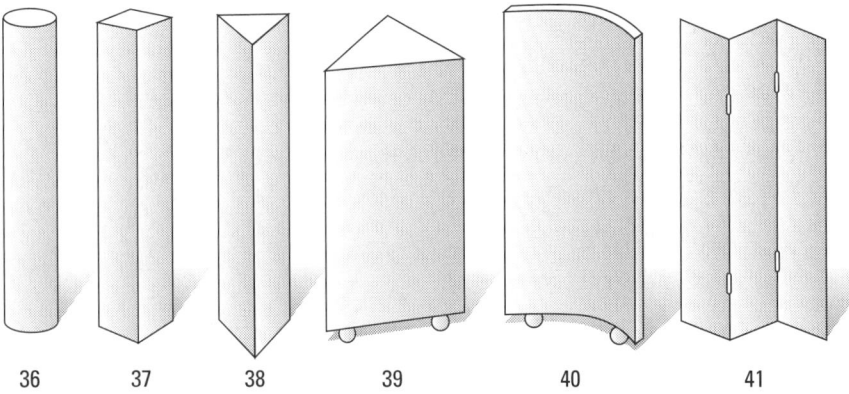

36   37   38   39   40   41

a. pylon

1a ○36 ○37 ○38
    ○39 ○40 ○41

b. screen

1b ○36 ○37 ○38
    ○39 ○40 ○41

c. pillar

1c ○36 ○37 ○38
    ○39 ○40 ○41

d. square pillar

1d ○36 ○37 ○38
    ○39 ○40 ○41

e. sweep

1e ○36 ○37 ○38
    ○39 ○40 ○41

f. periaktos

1f ○36 ○37 ○38
    ○39 ○40 ○41

PAGE TOTAL

Course No. _____ Date _____ Name _____

Select the correct answers, and fill in the bubbles with the corresponding numbers.

2. The standard backgrounds used to simulate interior and exterior walls are called (42) *cycs* (43) *flats* (44) *drops*.

3. The usual height for standard set units is (45) *7 feet* (46) *10 feet* (47) *14 feet*. For low-ceiling studios, it is (48) *8 feet* (49) *6 feet* (50) *12 feet*.

4. The continuous piece of canvas or muslin stretched along two, three, or even all four studio walls to form a uniform background is referred to as (51) *a drop* (52) *canvas backing* (53) *a cyclorama*.

5. In order to elevate scenery, properties, or action areas, we use (54) *periaktoi* (55) *platforms* (56) *pylons*.

6. The desk for the news anchor is considered a (57) *set prop* (58) *hand prop* (59) *set dressing*.

7. The pictures at the back of an interview set are a (60) *set prop* (61) *hand prop* (62) *set dressing*.

Chapter 13 — Production Environment: The Studio

8. From the rough floor plans shown below, select the one that most closely matches the simple sets shown on the facing page, and fill in the corresponding bubbles. *(Note that there are floor plans that do not match any of the set photos.)*

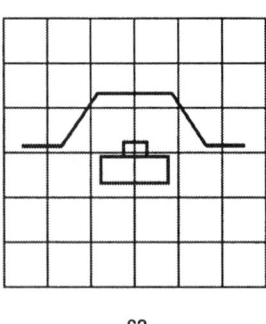
63

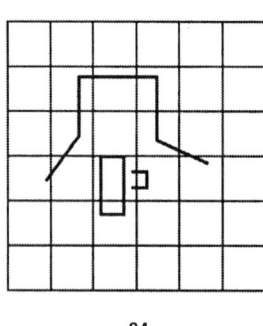
64

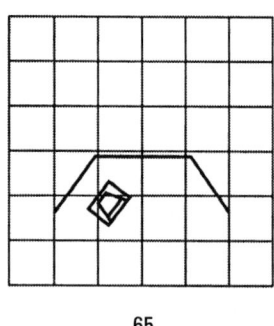
65

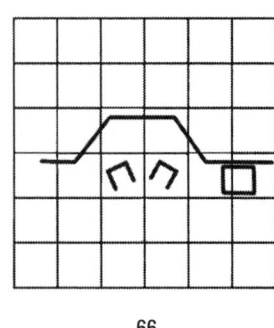
66

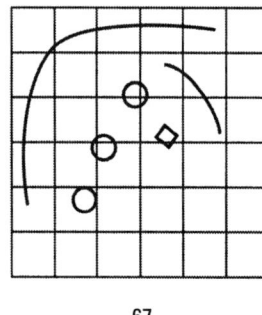
67

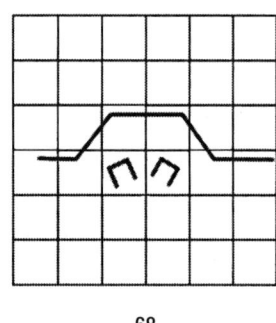
68

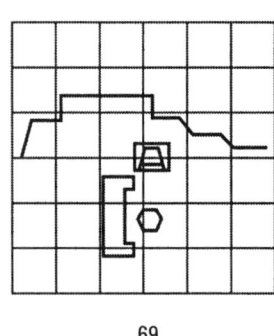
69

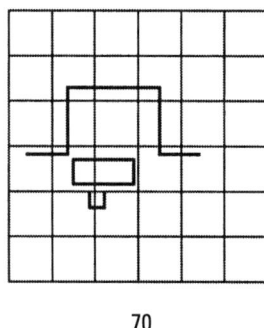
70

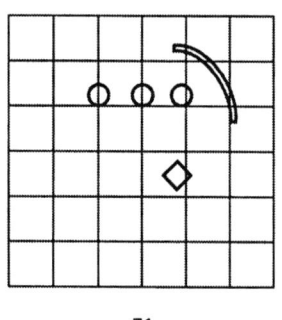
71

Course No. _____ Date _____ Name _____

a.

b.

c.

d.

e.

f.

Chapter 13 — *Production Environment: The Studio*

REVIEW QUIZ

*Mark the following statements as true or false by filling in the bubbles in the **T** (for true) or **F** (for false) column.*

1. Wagons can be used as platforms.

2. A small studio does not need a smooth floor because zoom lenses make dollying unnecessary.

3. All active furniture should be placed at least 6 to 8 feet from the background scenery.

4. A monitor is identical to a television set except that it has a sharper picture.

5. A well-functioning S.A. system makes a P.L. unnecessary.

6. In many video productions, props and set dressings are more important for indicating a certain style than the background flats.

7. Lashlines work well for softwall scenery but cannot be used for hardwall scenery.

8. The process message has a great influence on the set design.

9. An accurate control room clock makes the use of a stopwatch unnecessary.

10. The sound control (console) must be located as close to the director as possible.

SECTION TOTAL

Course No. _____ Date _____ Name _____

PROBLEM-SOLVING APPLICATIONS

1. You are asked by the new art professor about the feasibility of converting a classroom into a small video production studio. The classroom has no windows, normal doors, a 10-foot ceiling with normal fluorescent lighting, and a wood floor. What would you tell the art professor? Be specific.

2. You are asked by the company president why you need so many monitors in the control room. She tells you that she had consulted an electronics engineer, who told her that there are switching devices that let you preview various video sources on a single monitor. She quotes the engineer as saying that all you really need is two monitors. How, if at all, would you defend a multimonitor stack in the control room?

3. You are to evaluate the preliminary design for a new control room (see illustration below). What, if any, changes would you recommend? Why?

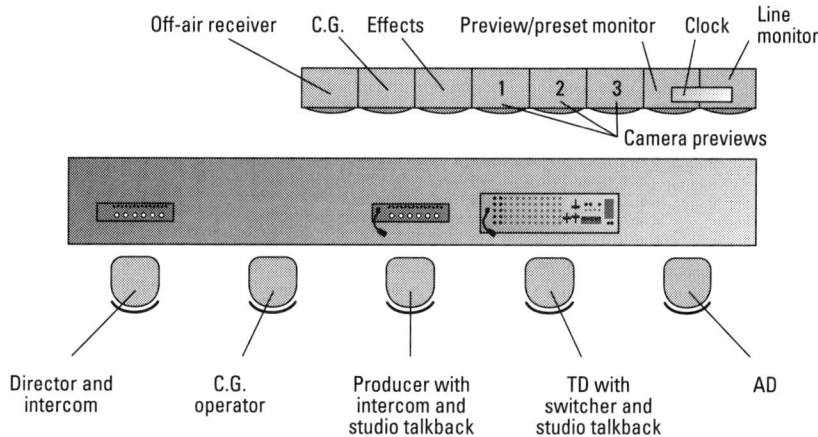

4. Using one of the forms on the following pages, draw a floor plan for a weekly interview show dealing with the art and media scene in your city. The host will interview guests from theater, the music scene, radio, and television. Include a detailed prop list.

5. Draw a floor plan for a morning news set. The news anchors consist of a woman and a man, and the news content is geared more toward local gossip than international politics.

Chapter 13 — Production Environment: The Studio

Course No. _____ Date _____ Name _____

Chapter 13 — *Production Environment: The Studio*

Property List

Course No. _____ Date _____ Name _____

14 Field Production and Synthetic Environments

REVIEW OF KEY TERMS

Match each term with its appropriate definition by filling in the corresponding bubble.

1. remote
2. field production
3. uplink truck
4. remote truck
5. ENG
6. EFP
7. remote survey
8. contact person
9. synthetic environment
10. virtual reality

A. Any video production that happens outside the studio.

B. A production of a large, scheduled event done for live transmission or live-on-tape recording.

C. The use of portable camcorders, lights, and sound equipment for the unplanned production of daily news stories. Usually done for live transmission or immediate postproduction.

D. Computer-simulated environment with which the user can interact and which can change according to the user's commands.

E. The vehicle that carries the control room, audio control, VTR section, video control section, and transmission equipment.

Chapter 14 — Field Production and Synthetic Environments

1. remote
2. field production
3. uplink truck
4. remote truck
5. ENG
6. EFP
7. remote survey
8. contact person
9. synthetic environment
10. virtual reality

F. Video production done outside the studio that is usually shot for postproduction (not live).

G. Electronically generated settings, either through chroma key or computer.

H. A person who is familiar with and who can facilitate access to the remote location and key people.

I. An inspection of the remote location by key production and engineering persons so that they can plan for the setup and use of production equipment.

J. Small truck that sends video and audio signals to a satellite.

Course No. _____ Date _____ Name _____

REVIEW OF REMOTES AND FIELD PRODUCTIONS

1. Evaluate the design for a new remote truck (see the following figure). The numbers in the sketch refer to the major pieces of control and operational equipment. From the list below, select the pieces of equipment that are indicated in the illustration (a through i), and fill in the bubbles with the corresponding numbers. Then determine which major equipment pieces were left out of the design, and fill in the appropriate bubbles for (18) *missing*.

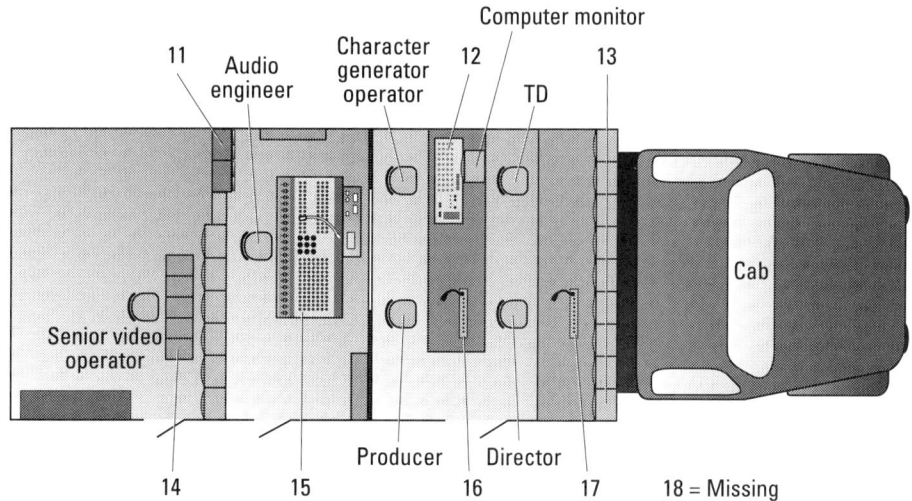

a. VTRs

b. CCUs

c. audio console

d. audio recording equipment

e. C.G.

Chapter 14 — Field Production and Synthetic Environments

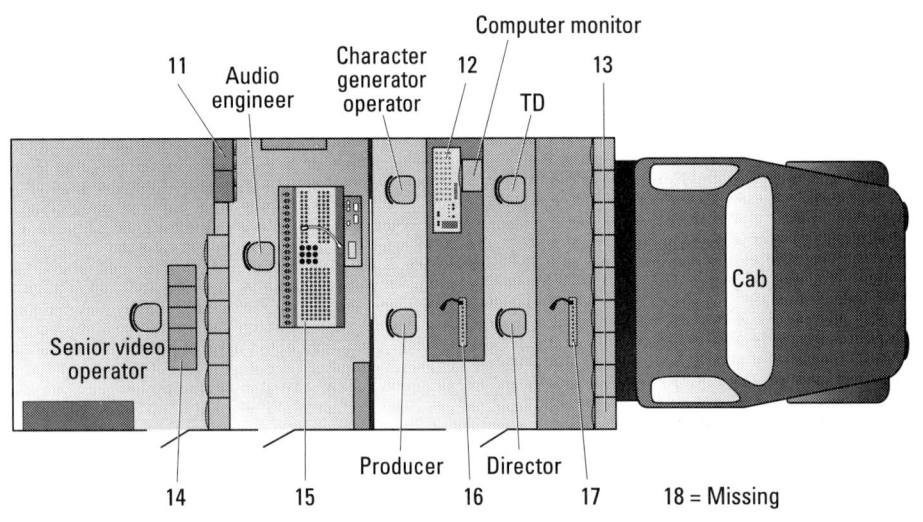

f. intercom controls

g. preview monitors

h. switcher

i. ESS

Select the correct answers, and fill in the bubbles with the corresponding numbers.

2. In case you are prevented from connecting the camera to the nearby remote truck by camera cable, you can transmit the camera signal by (19) *telephone wire* (20) *satellite downlink* (21) *microwave link* (22) *modem*.

3. The remote system least likely to use signal transmission equipment is (23) *ENG* (24) *EFP* (25) *big remotes*.

4. The field production least likely to need a remote survey is (26) *ENG* (27) *EFP* (28) *big remotes*.

Course No. _____ Date _____ Name _____

5. Analyze the following survey location sketch, and select the major production hazard from the list below. Fill in the bubble with the corresponding number.

5 ○ ○ ○ ○
   29 30 31 32

(29) *no room for cameras to maneuver*

(30) *plants interfering with good composition*

(31) *window causing lighting problems*

(32) *vector problems*

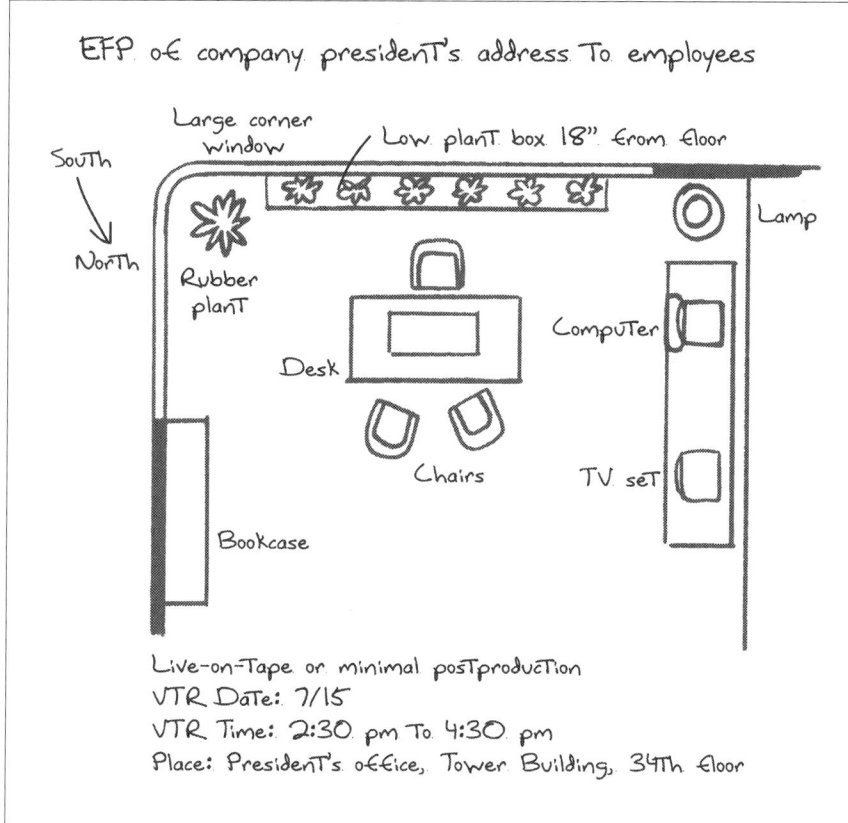

Chapter 14 — Field Production and Synthetic Environments

145

*Evaluate the equipment checklists for the three electronic field productions described below. Identify the **wrong equipment** or **items not needed** to take on location, and fill in the bubbles with the corresponding numbers.*

6. Taped interview of a media scholar in his hotel room for news item.

CHECKLIST

(33) *Betacam camcorder*
(34) *VTR*
(35) *portable lighting kit*
(36) *RCU*
(37) *shotgun mic and cable*
(38) *portable audio mixer*
(39) *videotape*
(40) *batteries*
(41) *preview monitors*

7. Taping of a brief dance number to recorded music in front of City Hall.

CHECKLIST

(42) *3 ENG/EFP cameras*
(43) *6 shotgun mics*
(44) *ESS*
(45) *3 VTRs*
(46) *3 RCUs and connecting cables*
(47) *large audio mixer*
(48) *P.A. audiotape playback system*
(49) *portable lighting kit*
(50) *C.G.*

8. Live stand-up traffic report from downtown during early-afternoon rush hour.

CHECKLIST

(51) *Betacam camcorder*
(52) *2 VTRs*
(53) *camera mic*
(54) *hand mic*
(55) *3 portable lighting kits*
(56) *audiotape recorder*
(57) *microwave transmission equipment*
(58) *I.F.B. intercom*
(59) *C.G.*

146   *Part IV — Talent and the Production Environment*

REVIEW OF SYNTHETIC ENVIRONMENTS

Select the correct answers, and fill in the bubbles with the corresponding numbers.

1. When using the chroma-key process, the background image must be provided by (60) *a live camera* (61) an *ESS system* (62) *any video source available*.

2. Virtual reality displays (63) *can* (64) *cannot* be integrated with analog video systems and (65) *can* (66) *cannot* show various points of view without a camera.

3. Interactive virtual reality means that you (67) *have a certain choice in what to see* (68) *can add to the available choices* (69) *have no choice in what video display to call up*.

Chapter 14 — Field Production and Synthetic Environments

REVIEW QUIZ

Mark the following statements as true or false by filling in the bubbles in the T (for true) or F (for false) column.

1. There is no difference between a lens-generated and a computer-generated environment.

2. Remote surveys are especially important for ENG.

3. You should do the survey for an outdoor remote during the time the actual production will take place.

4. A satellite news vehicle (SNV) is essential for transmitting a live ENG pickup from a remote location.

5. Satellite transmission is not possible unless the big remote truck has an uplink.

6. ENG and EFP need the same care of preproduction.

7. An I.F.B. intercom system is essential for successful ENG.

8. Contrary to ENG, EFP does not require keeping accurate field logs.

9. A well-equipped remote truck must contain a compact program control center.

10. All synthetic environments must originally be lens-generated.

SECTION TOTAL

Course No. _____ Date _____ Name _____

ZETTL'S VIDEO LAB 2.0 QUIZ

Click on the **lights** monitor and take the quiz on tape 9 **Field**. Click on the **audio** monitor and take the quizzes on tape 4 **Connectors** and tape 6 **Aesthetics**. Click on the **editing** monitor and take the quiz on tape 5 **Location Procedures**.

PROBLEM-SOLVING APPLICATIONS

1. To get a good overhead shot of a parade, you would like to place one of your cameras in the twentieth-floor window of a nearby hotel. Your TD informs you that the hotel has nothing against your renting the room for the day and setting up the camera, but will not allow any cable runs either inside or outside the hotel. What would you suggest?

2. To be "on the cutting edge" with his employees, the new company president would like to speak to them from a fantasy landscape, yet remain behind his familiar desk. Can you oblige the president's request? If so, how? If not, why not?

3. You are asked to assist in drawing up specifications for a new remote truck for Triple-I. What questions should you ask before making a recommendation? Why?

4. A new intern at Triple-I is very upset about your failure to request an adequate P.L. system for your next single-camera EFP project in the university student union. What would you tell the new intern? Why?

5. The new intern questions why you insist on keeping a field log, especially since the source tapes will be logged later anyway in preparation for postproduction editing. What is your reply? Why?

6. Conduct a detailed remote survey of two of the following events: a local school board meeting; a rock concert in a city park; a modern dance performance in front of City Hall; an interview with a university president in his or her office; the gala opening of a new computer store; a wedding in a local church; a basketball game in a high-school or college gym.

Chapter 14 — Field Production and Synthetic Environments

Scale: 1/4" = 1'

Property List

Scale: 1/4" = 1'

Property List

Scale: 1/4" = 1'

Property List

Scale: 1/4" = 1'

Property List

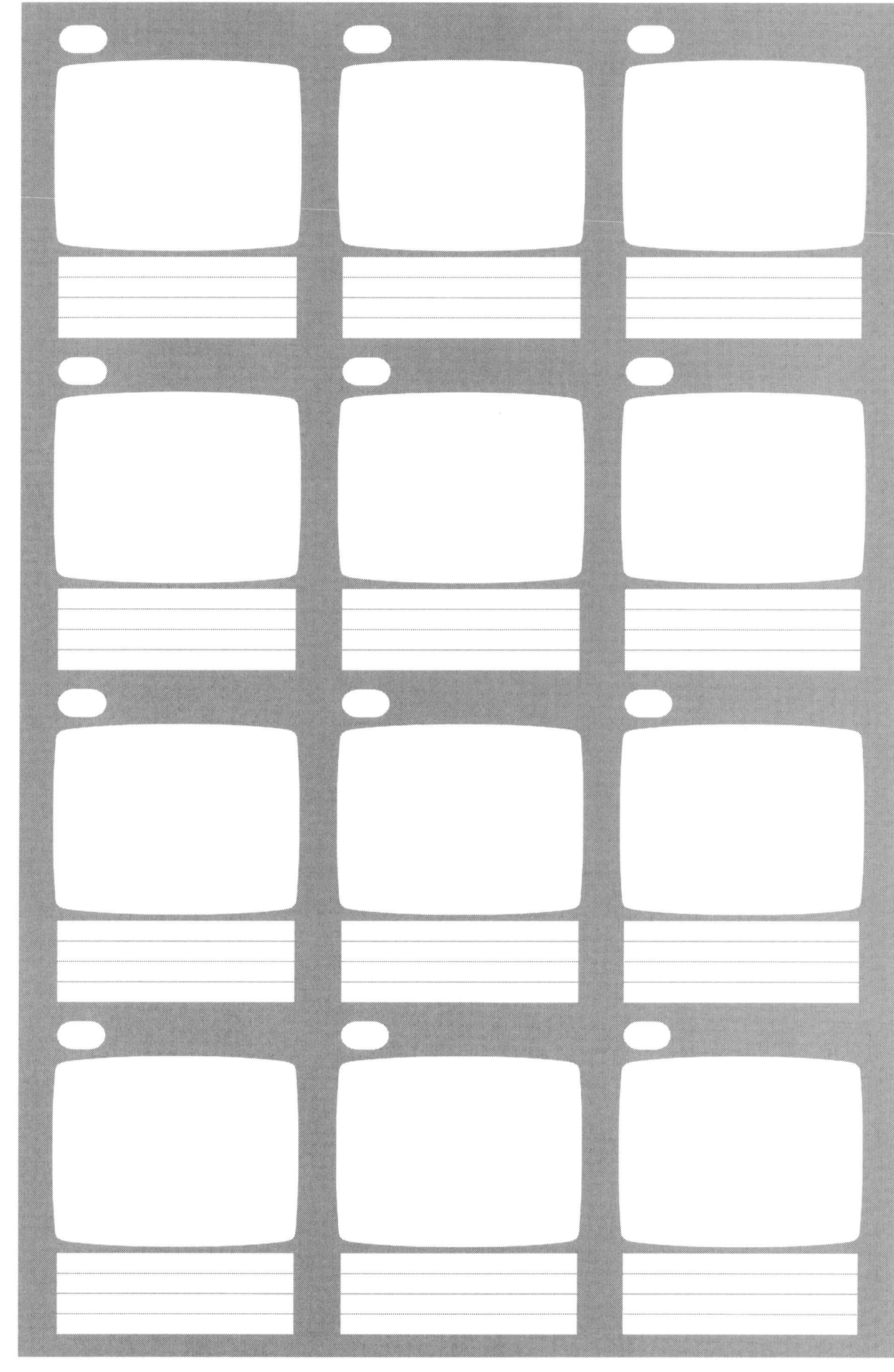

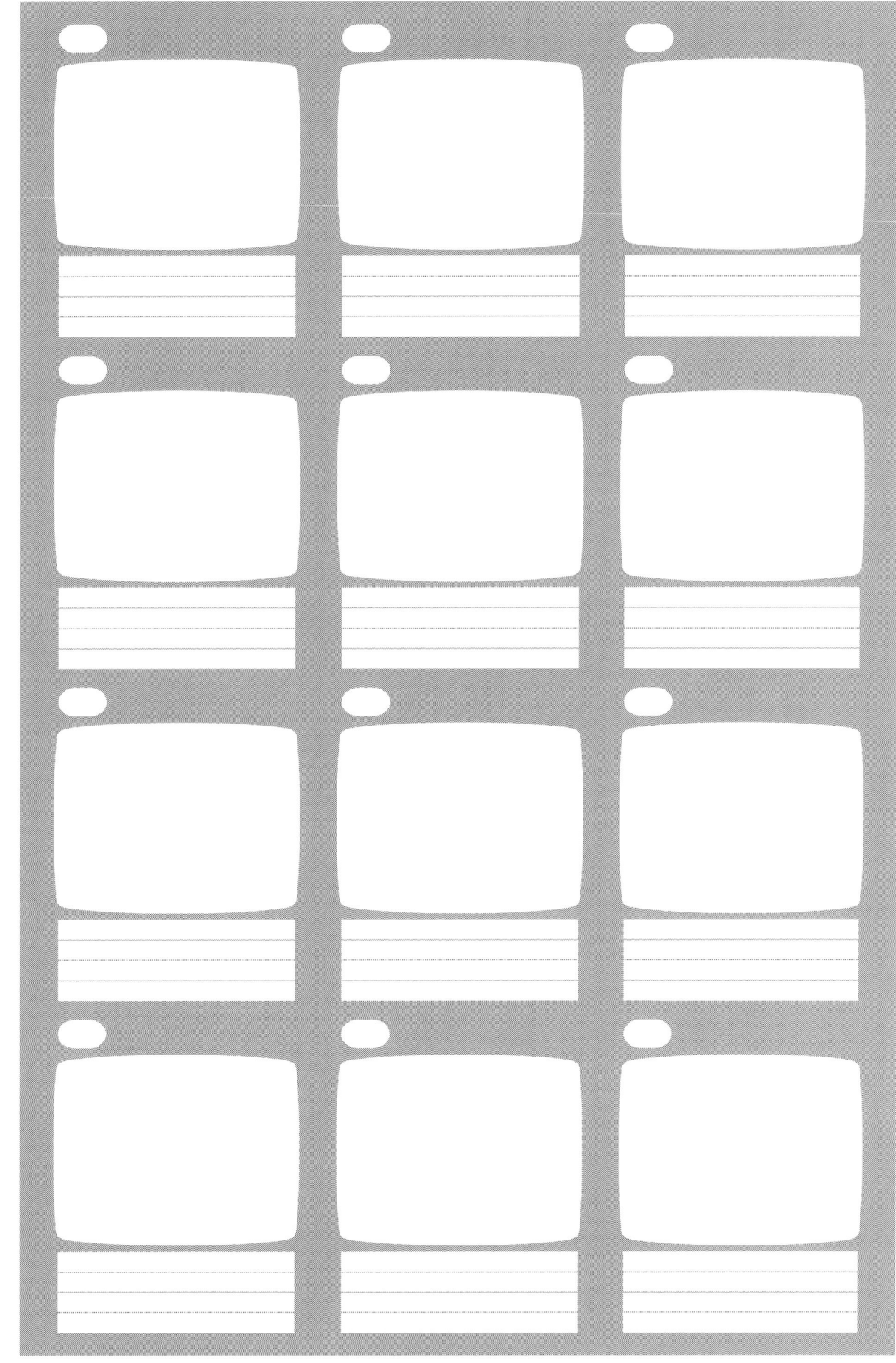

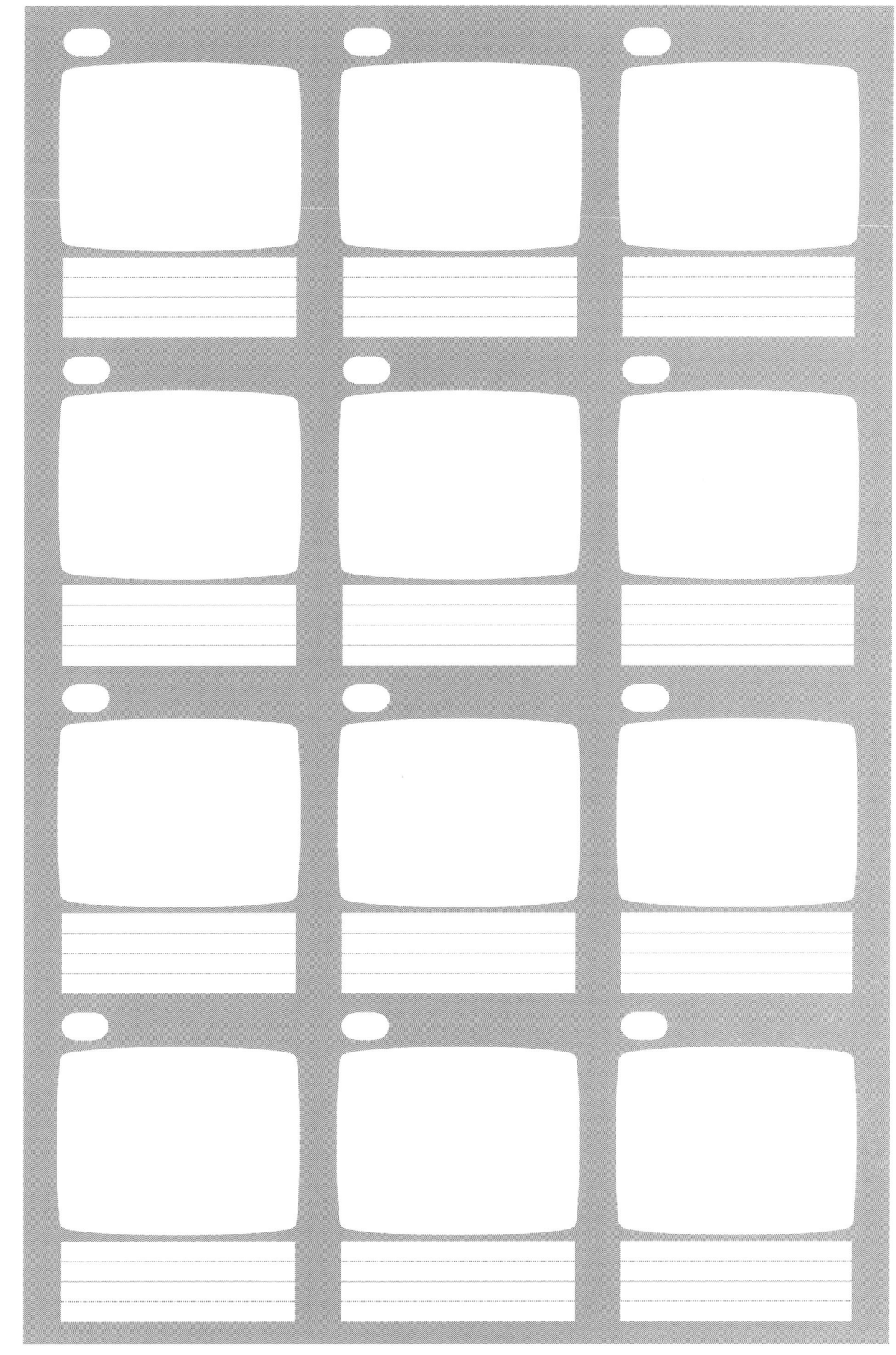

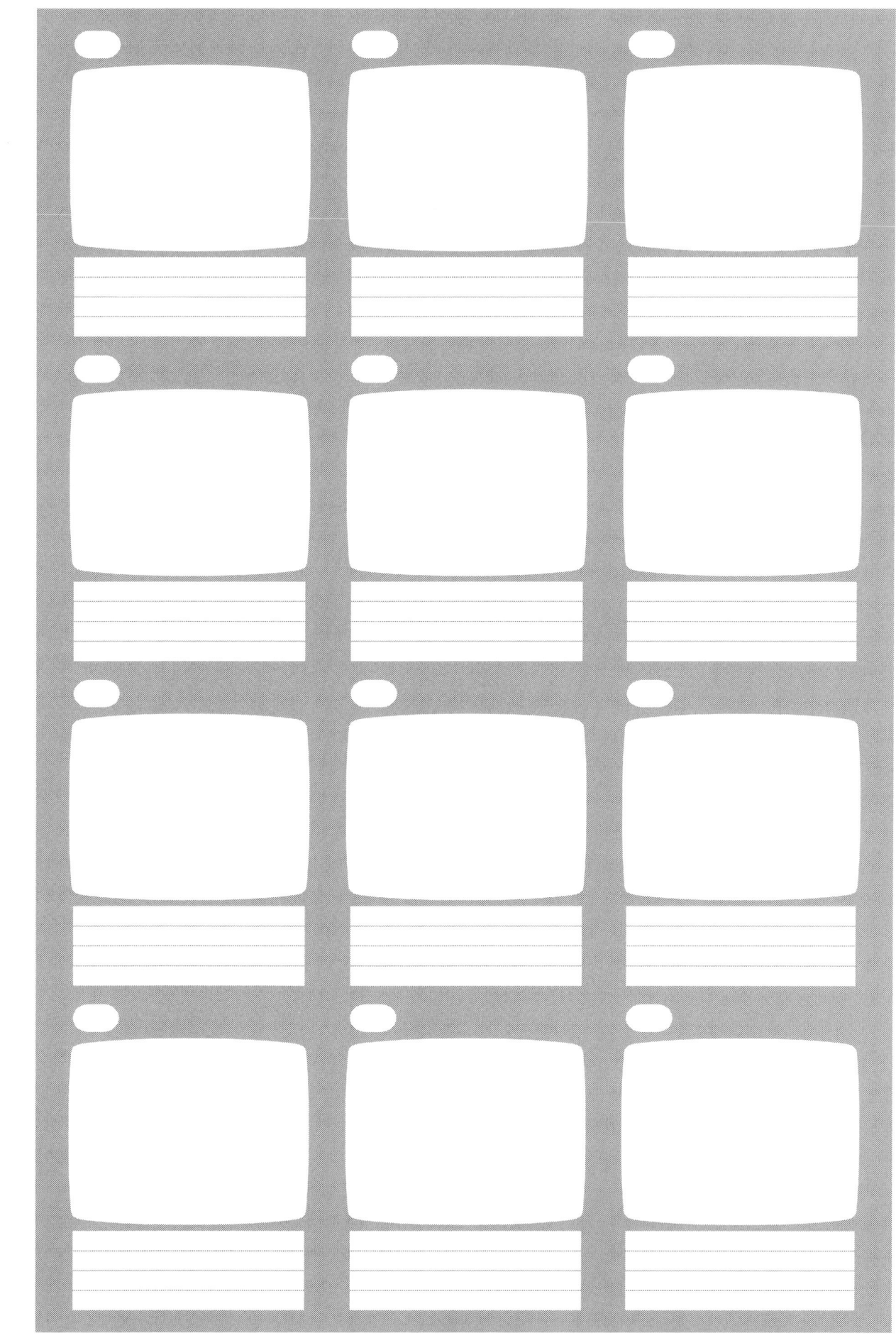

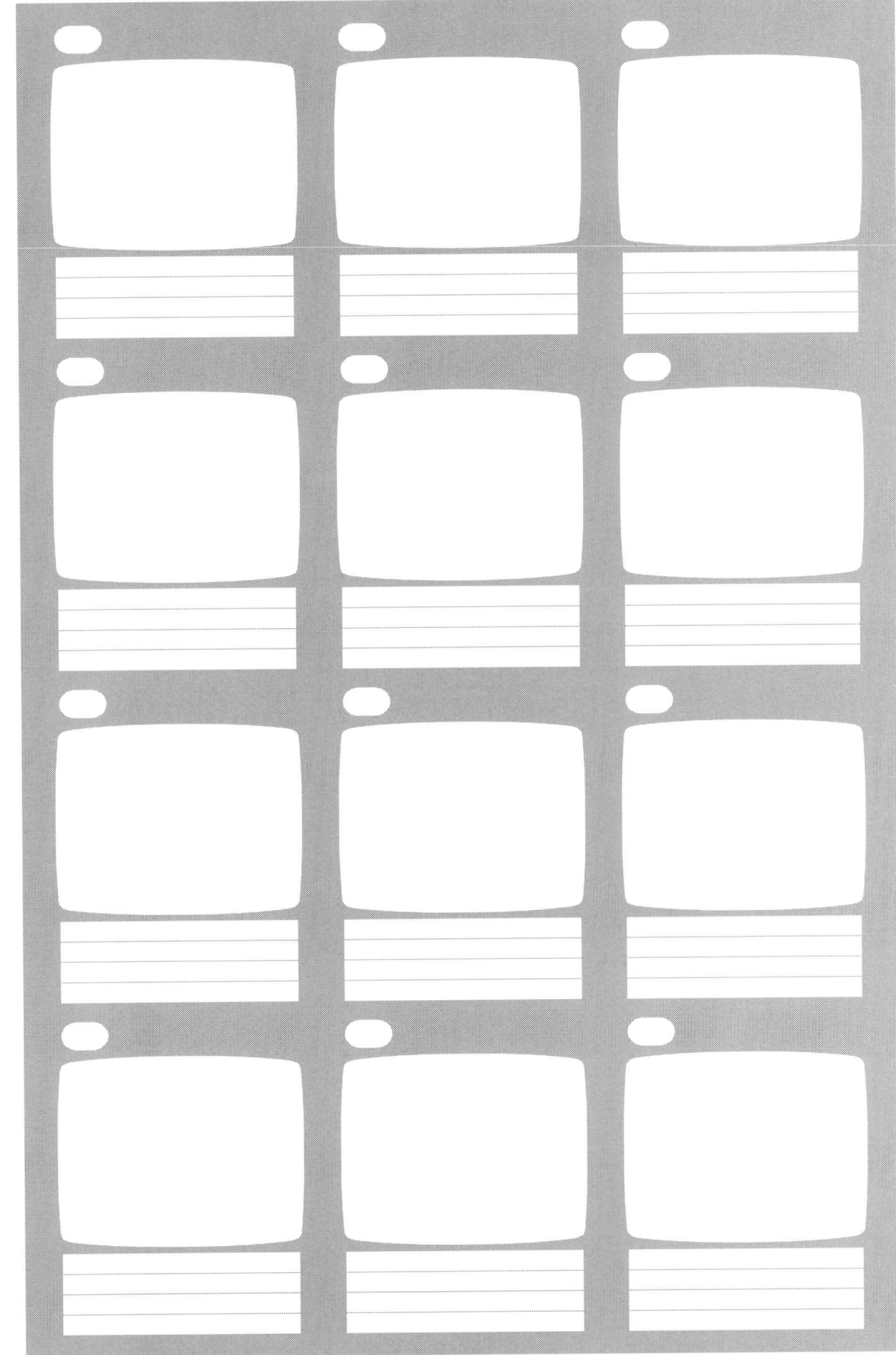